Manual of Water Supply Practices

Backflow Prevention and Cross-Connection Control
Recommended Practices

Fourth Edition

American Water Works Association

Manual of Water Supply Practices—M14, Fourth Edition

Backflow Prevention and Cross-Connection Control: Recommended Practices

Copyright ©1973, 1989, 2004, 2015 American Water Works Association

All rights reserved. No part of this publication may be reproduced or transmitted in any form or by any means, electronic or mechanical, including photocopy, recording, or any information or retrieval system, except in the form of brief excerpts or quotations for review purposes, without the written permission of the Publisher.

Disclaimer
The authors, contributors, editors, and publisher do not assume responsibility for the validity of the content or any consequences of their use. In no event will AWWA be liable for direct, indirect, special, incidental, or consequential damages arising out of the use of information presented in this book. In particular, AWWA will not be responsible for any costs, including, but not limited to, those incurred as a result of lost revenue. In no event shall AWWA's liability exceed the amount paid for the purchase of this book.

If you find errors in this manual, please email books@awwa.org. Possible errata will be posted at www.awwa.org/resources-tools/resource-development-groups/manuals-program.aspx.

Project Manager/Senior Technical Editor: Melissa Valentine
Senior Manuals Specialist: Molly Beach
Senior Production Editor: Cheryl Armstrong

Library of Congress Cataloging-in-Publication Data
Asay, Stuart F., author.
  Backflow prevention and cross connection control : recommended practices / Stuart F. Asay. -- 4th edition.
     pages cm. -- (Manual of water supply practices ; M14)
  Revision of: Recommended practice for backflow prevention and cross-connection control, 3rd ed., 2004.
  Includes bibliographical references and index.
  ISBN 978-1-62576-045-6 (alk. paper)
  1. Backsiphonage (Plumbing)--Prevention. 2. Cross-connections (Plumbing) III. Title. IV. Title: Recommended practice for backflow prevention and cross-connection control. V. Series: AWWA manual ; M14.
  TH6523.A87 2015
  696'.1--dc23
                              2014037357

Printed in the United States of America

ISBN: 978-1-62576-045-6   eISBN: 978-1-61300-309-1

This AWWA content is the product of thousands of hours of work by your fellow water professionals. Revenue from the sales of this AWWA material supports ongoing product development. Unauthorized distribution, either electronic or photocopied, is illegal and hinders AWWA's mission to support the water community.

American Water Works Association

American Water Works Association
6666 West Quincy Avenue
Denver, CO 80235-3098
awwa.org

Contents

List of Figures, v

List of Tables, vii

Preface, ix

Acknowledgments, xi

Dedication, xiii

Metric Conversions, xv

Chapter 1 Introduction 1
> Purpose of Manual, 1
> Responsibilities, 2
> Health Aspects, 5
> Legal Aspects, 7

Chapter 2 Backflow Prevention Principles 11
> Basic Hydraulics, 11
> Types of Backflow, 13
> Assessing Degrees of Hazard, 18
> Assessing Other Risk Factors, 20
> Assessing the Effectiveness of Assemblies and Devices, 21

Chapter 3 Program Administration 23
> Developing a Cross-Connection Control Program, 24
> Types of Programs, 25
> Management Programs, 28
> Documentation, 33
> Human Resources, 33
> Program Administration, 35

Chapter 4 Backflow Prevention Assembly Application, Installation, and Maintenance 41
> Means of Preventing Backflow, 42
> Backflow Prevention Devices, 42
> Method for Controlling Backflow, 59
> Field Testing, 61
> Reference, 64

Chapter 5 Conducting a Cross-Connection Control Survey 65
> Authority and Responsibilities, 66
> Purpose of a Cross-Connection Control Survey, 66
> Assessing the Degree of Hazard, 67
> Survey Considerations and Concepts, 67
> Conducting the Survey, 69
> Reference, 74

Chapter 6 Sample Hazards and Proper Protection 75
> Typical Hazards, 76
> Hazards Posed by a Water Supplier, 84
> Protection for Specific Customers, 89
> References, 91

Appendix A   Example Notices and Letters, 93

Appendix B   Testing Procedures or Methods, 99

Appendix C   Industry Resources, 181

Appendix D   Incidents Tables, 183

Glossary, 187

Index, 193

List of AWWA Manuals, 203

Figures

1-1 Examples of backflow prevention equipment locations, 4

2-1 Example of barometric loop in a piping configuration, 12
2-2 Backsiphonage backflow due to high rate of water withdrawal, 14
2-3 Backsiphonage backflow caused by reduced pressure on suction side of booster pump, 15
2-4 Backsiphonage backflow caused by shutdown of water system, 15
2-5 Backpressure backflow caused by carbon dioxide cylinder, 17
2-6 Backpressure backflow caused by pumping system, 18

4-1 Common symbols used for backflow prevention devices, 43
4-2 Dual check device, 44
4-3 Dual check device with atmospheric port, 45
4-4 Atmospheric vacuum breaker, 45
4-5 Hose connection vacuum breaker, 46
4-6 Pressure vacuum breaker assembly, normal flow condition, 48
4-7 Pressure vacuum breaker assembly, backsiphonage condition, 48
4-8 Spill-resistant vacuum breaker, normal flow and backsiphonage conditions, 50
4-9 Check valves open, permitting flow, 51
4-10 Backpressure, both check valves closed, 51
4-11 Negative supply pressure, check valves closed, 51
4-12 Typical double check valve assembly applications, 52
4-13 Reduced-pressure principle backflow prevention assembly, both check valves open and the differential relief valve closed, 54
4-14 Both check valves closed and the differential pressure relief valve open can be used for service protection or internal protection, 54
4-15 Backpressure: both check valves closed and the differential pressure relief valve closed, 54
4-16 Backsiphonage: both check valves closed and the differential pressure relief valve open, 55
4-17 Typical reduced-pressure principle backflow prevention application, 55
4-18 Double check detector backflow prevention assembly, 57
4-19 Type II double check detector backflow prevention assembly, 57
4-20 Reduced-pressure principle detector backflow prevention assembly, 58
4-21 Type II reduced-pressure principle detector backflow prevention assembly, 58
4-22 AG on tank, 59
4-23 AG on lavatory, 60
4-24 Typical AG applications, 60
4-25 Additional typical AG applications, 61

6-1 Cross-connection control, water treatment plants, 87
6-2 Service-containment and area-isolation water treatment plants, 88

B-1 Double check valve assembly test, 101
B-2 Double check valve assembly test—back pressure condition, 102
B-3 DCVA—Step #1, 103

BACKFLOW PREVENTION AND CROSS-CONNECTION CONTROL

B-4   DCVA—Step #2, 104
B-5   DCVA—Step #3 and DCVA Step #3A, 105
B-6   Pressure vacuum breaker, 107
B-7   PVB—Step #1, 108
B-8   PVB—Step #2, 109
B-9   PVB—Step #3, 110
B-10 RPZ—Step #1A, 112
B-11 RPZ—Step #2, 113
B-12 RPZ—Step #3, 114
B-13 RPZ—Step #4, 115
B-14 RPZ—Step #5, 116
B-15 Spill-resistant pressure vacuum breaker, 117
B-16 Step 1, 118
B-17 Step 2, 119
B-18 Step 3, 120
B-19 Sample test report form, 124
B-20 Ground wire installation, 127
B-21 Major component parts of five-valve differential pressure gauge test equipment, 139
B-22 Differential pressure gauge test kit, five-valve model, 140
B-23 Differential pressure gauge showing hose connections to test the components of an RPBA, 140
B-24 Differential pressure gauge showing hose connections to test the #1 check valve of a DCVA, 141
B-25 Differential pressure gauge showing hose connections to test a PVBA air inlet, 141
B-26 Differential pressure gauge showing hose connections to test a PVBA check valve, 142
B-27 Differential pressure gauge showing hose connections to test both the check valve and air inlet of an SVBA, 142
B-28 Test 1—CV#1, 171
B-29 Test 3—CV#2, 172
B-30 Two-valve differential pressure test kit, 176
B-31 Two-valve test kit, 176
B-32 Three-valve test kit, 178
B-33 Reduced-pressure field test with three-valve test kit, 179

Tables

2-1 Means of backflow prevention in the United States, 22
2-2 Selection guide for backflow preventers in Canada, 22

6-1 Recommended protection for solar domestic hot-water systems, 84
6-2 Recommended protection at fixtures and equipment found in water treatment plants, 86
6-3 For service protection (containment), 89
6-4 Containment protection, 90
6-5 Typical backflow prevention devices, 90
6-6 Irrigation and hose connection protection, 91

B-1 Safety-related publications, 129
B-2 RPBA/RPDA test reporting, 133
B-3 DCVA/DCDA test reporting, 134
B-4 PVBA/SVBA test reporting, 137
B-5 Approved minimum test result values, 137

This page intentionally blank.

Preface

This is the fourth edition of the AWWA Manual M14, *Backflow Prevention and Cross-Connection Control: Recommended Practices*. It provides both technical and general information to aid in the development, implementation, and management of a cross-connection control and backflow prevention program, and an understanding of backflow prevention and cross-connection control concepts. This manual is a review of recommended practice. It is not an AWWA standard calling for compliance with certain requirements. It is intended for use by water suppliers and municipalities of all sizes, whether as a reference book or a textbook for those not familiar with backflow prevention and cross-connection control.

This manual reviews regulatory provisions established to protect the potable water supply. To achieve this goal, products and measures are discussed to assist in the determination of controlling hazardous cross-connections. For fundamental knowledge and a thorough understanding, this entire manual should be carefully studied. Readers will also find the manual a useful source of information when assistance is needed with specific or unusual connections to the potable water supply.

This fourth edition of M14 includes updates to regulatory concerns and products that protect the water supply, and new material on establishing programs to control cross-connections, including surveying of piping systems to identify and monitor such connections.

This page intentionally blank.

Acknowledgments

The AWWA Technical and Educational Council, the Distribution Plant Operations Division, and the Cross-Connection Control committee gratefully acknowledge the contributions made by those volunteers who drafted, edited, and provided the significant and critical commentary essential to updating M14. The Technical Review Board members dedicated numerous hours in the final stages of preparation of this edition to ensure the overall technical quality, consistency, and accuracy of the manual.

Technical Review Board Members

Stuart F. Asay, Backflow Prevention Institute IAPMO, Westminster, Colo.
Joseph A. Cotruvo, Joseph Cotruvo & Associates LLC, Washington, D.C.
Dawn M. Flancher, American Water Works Association, Denver, Colo.
Aneta King, Halton Regional Muncipality, Oakville, ON, Canada
Randolph J. Pankiewicz, Illinois American Water, Belleville, Ill.

Contributors to the 4th edition

Rand H. Ackroyd, Deceased, Rand Technical Consulting LLC, Newburyport, Mass.
Stuart F. Asay, Backflow Prevention Institute IAPMO, Westminster, Colo.
Roland Asp, National Fire Sprinkler Association Inc., Patterson, N.Y.
Nick Azmo, Azmo Mechanical Inc., Piscataway, N.J.
Lou Allyn Byus, Retired, Jacksonville, Ill.
Dave Bries, City of Montrose, Montrose, Colo.
Nicole Charlton, Philadelphia Water Department, Philadelphia, Pa.
Sean Cleary, Backflow Prevention Institute IAPMO, Scranton, Pa.
Richard A. Coates, Retired, Miami, Fla.
Alicia A. Connelly, City of Norfolk, Norfolk, Va.
Daniel Eisenhauer, Backflow Solutions Inc., Alsip, Ill.
Dylan Gerlack, EPCOR Water Services, Edmonton, AB, Canada
Steve Gould, Peterborough Utilities Commission, Peterborough, ON, Canada
John F. Graham, California Water Service Company, Chico, Calif.
James W. Green, City of Calgary, Calgary, AB, Canada
Byron Hardin, Hardin and Associates Consulting LLC, Coppell, Texas
John F. Higdon, Conbraco Industries, Matthews, N.C.
James Holeva, retired, Fall River, Mass.
Robert Hunter, City of Norfolk, Norfolk, Va.
Alissa Kantola, Val-Matic Valves & Manufacturing Corp., Elmhurst, Ill.
Aneta King, Halton Regional Municipality, Oakville, ON, Canada
Mark S. Kneibel, Hydro Designs Inc., Jenison, Mich.
Diane Meyer, Val-Matic Valves & Manufacturing Corp., Elmhurst, Ill.
Vince Monks, Louisville Water Company, Louisville, Ky.
Dan O'Lone, USEPA, Atlanta, Ga.
John E. Ralston, Louisville Water Company, Louisville, Ky.
Paul H. Schwartz, University of Southern California FCCCHR, Los Angeles, Calif.
Pauli Undesser, Water Quality Association, Lisle, Ill.
Barry Walter, Deceased, Hydro Designs Inc., Chester, Md.

Karl Wiegand, Globe Fire Sprinkler Corp., Standish, Mich.
Michelle J. Williams, City of Norfolk, Norfolk, Va.

Agency Contributors to the 4th edition

New England Water Works Association, a Section of AWWA
AWWA Pacific Northwest Section
The American Society of Sanitary Engineering (ASSE)
University of Florida, Center for Training, Research, and Education for Environmental Occupations (UF TREEO)

Dedication

The 4th edition of AWWA's Manual M14, *Backflow Prevention and Cross-Connection Control: Recommended Practices*, is dedicated to Rand Ackroyd and Barry Walter. During the revision of this manual edition, the industry lost two valuable professionals that were vital resources and contributors who eagerly shared their collective wisdom.

Rand Ackroyd

Rand Ackroyd became an active member of the M14 committee in the early 1980s. His experience as vice president of engineering for Watts Regulator Co. brought a wealth of technical information regarding backflow preventers, their capabilities, and proper applications. After leaving Watts, he continued his active involvement as a go-to resource for nearly every aspect of backflow prevention.

Barry Walter

Barry Walter will be remembered as a professional who was passionate about his work in the backflow prevention industry. He derived great satisfaction from teaching a group about the subject and from participants gaining an understanding of the important health and safety aspects of the subject. While working on this edition of the manual, he was instrumental in coordinating several chapter teams, working toward consensus, and striving to achieve the best reference manual on the subject.

These valuable committee members will be missed but never forgotten.

This page intentionally blank.

Metric Conversions

Linear Measurement

inch (in.)	×	25.4	=	millimeters (mm)
inch (in.)	×	2.54	=	centimeters (cm)
foot (ft)	×	304.8	=	millimeters (mm)
foot (ft)	×	30.48	=	centimeters (cm)
foot (ft)	×	0.3048	=	meters (m)
yard (yd)	×	0.9144	=	meters (m)
mile (mi)	×	1,609.3	=	meters (m)
mile (mi)	×	1.6093	=	kilometers (km)
millimeter (mm)	×	0.03937	=	inches (in.)
centimeter (cm)	×	0.3937	=	inches (in.)
meter (m)	×	39.3701	=	inches (in.)
meter (m)	×	3.2808	=	ft (ft)
meter (m)	×	1.0936	=	yards (yd)
kilometer (km)	×	0.6214	=	miles (mi)

Area Measurement

square meter (m^2)	×	10,000	=	square centimeters (cm^2)
hectare (ha)	×	10,000	=	square meters (m^2)
square inch ($in.^2$)	×	6.4516	=	square centimeters (cm^2)
square foot (ft^2)	×	0.092903	=	square meters (m^2)
square yard (yd^2)	×	0.8361	=	square meters (m^2)
acre	×	0.004047	=	square kilometers (km^2)
acre	×	0.4047	=	hectares (ha)
square mile (mi^2)	×	2.59	=	square kilometers (km^2)
square centimeter (cm^2)	×	0.16	=	square inches ($in.^2$)
square meters (m^2)	×	10.7639	=	square ft (ft^2)
square meters (m^2)	×	1.1960	=	square yards (yd^2)
hectare (ha)	×	2.471	=	acres
square kilometer (km^2)	×	247.1054	=	acres
square kilometer (km^2)	×	0.3861	=	square miles (mi^2)

Volume Measurement

cubic inch ($in.^3$)	×	16.3871	=	cubic centimeters (cm^3)
cubic foot (ft^3)	×	28,317	=	cubic centimeters (cm^3)
cubic foot (ft^3)	×	0.028317	=	cubic meters (m^3)
cubic foot (ft^3)	×	28.317	=	liters (L)
cubic yard (yd^3)	×	0.7646	=	cubic meters (m^3)

This page intentionally blank.

AWWA MANUAL

M14

Chapter **1**

Introduction

For millennia, people have been concerned with obtaining and maintaining pure and safe water supplies. Archeological studies reveal that as early as 3000 BC, the ancient Egyptian State had a government official who was required to inspect the country's water supply every 10 days. With the widespread use of water closets in the 1800s came direct cross-connections with water mains. This brought into focus the problem that, as one nineteenth century authority stated, "foul matters may get into the pipes."[*] Currently, many government and industry professionals are aware of the need to prevent contamination of potable water supplies through cross-connections. However, the water supplier goals and levels of involvement may vary.

PURPOSE OF MANUAL

This manual provides guidance to all professionals working with the potable water supply on the recommended procedures and practices for developing, operating, and maintaining an efficient and effective cross-connection control program. The manual also provides insight into the basic areas that should be addressed to ensure that public water system connections are made safely; that those connections will be operated and maintained to ensure water quality; and that public water suppliers have the basic knowledge needed to assist in this effort. The purpose of any such program is to reduce the risk of contamination or pollution of the public water system.

A cross-connection is an actual or potential connection between any part of a potable water system and an environment that would allow substances to enter the potable water system. Those substances could include gases, liquids, or solids, such as chemicals, water products, steam, water from other sources (potable or nonpotable), and any matter that may change the color or taste of water or add odor to water.

---

[*] A.J. Keenan, C.S.I., B.C. Section AWWA Cross-Connection Control, September 1977.

RESPONSIBILITIES

The United States Safe Drinking Water Act (SDWA) became law in 1974. The purpose of the act is to protect public health by regulating all public drinking water supplies in the United States. SDWA was amended in 1986 and again in 1996. As amended, it requires protection of the public drinking water supply and its sources: both surface water and ground water. SDWA does not, however, regulate private wells serving fewer than 25 individuals and it also does not regulate systems having fewer than 15 service connections.

SDWA authorizes the United States Environmental Protection Agency (USEPA) to set national health-based standards for public drinking water. These standards have been established to protect against naturally occurring and man-made contaminants that may be found in our drinking water supply. Together the USEPA, state regulatory agencies, and water suppliers work to make sure these standards are monitored and followed.

In Canada, provincial governments have jurisdiction over the public health aspects of the drinking water supply. Local governments within a province (e.g., regional districts and municipalities), with the authority of the province, may impose other regulations or more stringent regulations not in conflict with provincial regulations.

Because there is a difference between the authority of the United States federal government and that of Canada, and between the different states and provinces, the following discussion, although referring to "federal and state," illustrates the different regulations governing a public water supplier.

For US water utilities (public water suppliers), SDWA regulations govern PWSs. SDWA (see 42 U.S.C. 300f(4)(A)) states: "The term *public water system* [PWS] means a system for the provision to the public of water for human consumption through pipes or other constructed conveyances, if such system has at least fifteen service connections or regularly serves at least twenty-five individuals."

The public water system includes any collection, treatment, storage, and distribution facilities under control of the operator of such system.

SDWA states that water suppliers are only responsible for the water quality delivered to the water consumer's service connection. In many jurisdictions, this is commonly referred to as *point of entry* or *point of service*. The water supplier is not responsible for contaminants and/or pollutants that are added to the potable water by any circumstances under the control of the consumer beyond the public water supply water point of entry.

Currently, SDWA provides that the federal government may grant the state or local governments primacy for the administration and enforcement of the federal drinking water rules and regulations. Agencies that have been granted primary enforcement responsibilities must adopt drinking water regulations that are at least as stringent as the current federal drinking water rules and regulations.

In addition to adopting the federal drinking water rules and regulations, primacy agencies may adopt additional or more stringent drinking water rules or regulations as long as the rules or regulations are not in conflict with SDWA and/or other federal rules or regulations. Thus, in states that have primacy, cross-connection control rules are state adopted. Similarly, local government such as counties and cities, with the authority of the state, may also adopt additional or more stringent rules and regulations as long as the counties and cities are not in conflict with the state law or regulations.

Both federal and state/provincial governments regulate the public health aspects of drinking water in order to protect the health, safety, and welfare of the water consumer.

From the point of service/entry, federal, state, and local responsibilities to protect the health, safety, or welfare of the users of water are under the jurisdiction of agencies other than those regulating PWSs, and include, but are not limited to, the following:

- Local plumbing and building officials who are responsible for enforcing all provisions of applicable plumbing and building codes relative to the installation, alteration, repair maintenance, or operation of all plumbing system devices and equipment including cross-connections on all new construction or any project that has plumbing or any building that has a construction, plumbing, or building permit open. In such cases, the local code official will make all the required inspections or they may accept reports of inspection by approved agencies or individuals. Plumbing codes provide for point-of-use backflow protection for potable water systems. The local code officials usually have limited or no jurisdiction over any pre-existing structures.

- Fire marshals who are responsible for regulating fire protection systems (e.g., fire sprinkler systems) downstream of the potable water system supply connection entering the premises.

- Safety inspectors (Occupational Safety and Health Administration [OSHA]; Workers' Compensation Board [WCB] [Canada]; Mine Safety and Health Administrators [MSHA]) who are responsible for inspecting potable water systems (plumbing) for workers' safety.

- Health officials who are responsible for inspecting restaurants and other food preparation facilities (e.g., dairies), health care facilities (e.g., nursing homes), etc.

- Agricultural inspectors who are responsible for the safe handling of chemicals (e.g., pesticides) used in growing and processing agricultural products.

These agencies have jurisdiction over work done on the customer's premises. Most have regulations that involve cross-connection control, and these different regulations may be in conflict with the procedures for cross-connection control recommended in this manual. The authority of these agencies over the water supplier's customers may be continuing or may be limited by the issuance of a final permit (e.g., for building occupancy) (see Figure 1-1).

The implementation of a program for the effective control of cross-connections requires the cooperation of the water supplier, the primacy agency, plumbing/building officials, plumbers, the water consumer, and the backflow prevention assembly tester. Each has specific responsibilities and each must carry out their responsibilities in order to prevent pollution or contamination of the PWS.

Much confusion about cross-connection control exists due to a misunderstanding between many water suppliers, property owners, and code officials that under SDWA, the water suppliers are responsible for water quality to the last free-flowing tap. This confusion may result from some federal rules/regulations requiring water suppliers to monitor certain water quality parameters, such as maximum contaminant level (MCL) violations or necessary action levels for treatment techniques, which are measured at the tap and are reflective of the corrosivity of the water being supplied. Additional information can be obtained from the local authority having jurisdiction. However, this does not impose a responsibility on the water supplier for regulating plumbing. As stated in SDWA, "Maximum contaminant level means the maximum permissible level of a contaminant in water which is delivered to any user of a public water system."

Cross-connection control regulations provide water utilities a legal basis for reviewing water users for actual or potential cross-connections. More importantly, they impose requirements that adequately protect the PWS whenever a potential hazard is discovered.

4   BACKFLOW PREVENTION AND CROSS-CONNECTION CONTROL

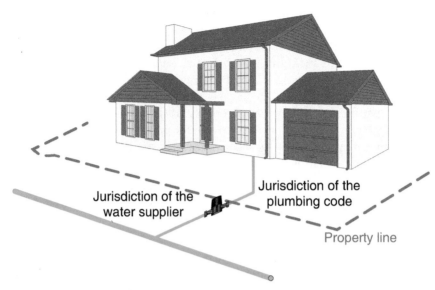

A. Location on private side of property line

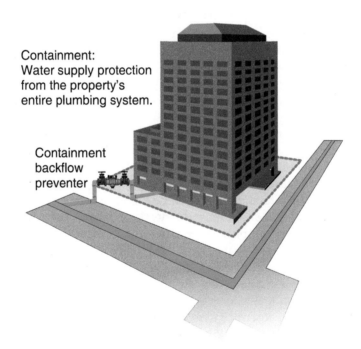

B. Location on public entity side of property line

Figure 1-1   Examples of backflow prevention equipment locations
Courtesy of Stuart F. Asay

HEALTH ASPECTS

Protection of drinking water for public health emphasizes preventing contamination. A multiple-barrier approach is used from the source to the tap. The following are major barriers established for PWSs:

- **Sources of supply:** Prevent human contaminants, such as pathogens (e.g., viruses and bacteria) or chemicals from entering the water supply through watershed control and wellhead protection programs.
- **Treatment techniques:** Remove or reduce natural and human contaminants to comply with maximum contaminant levels established by regulations, or otherwise provided by the system.
- **Chlorination:** Maintain chlorine residual in the water supply to control microbiological quality.
- **Storage:** Provide covered storage and prevent microbiological contamination through openings in reservoirs.
- **Distribution:** Comply with installation and material standards and provide minimum operating pressures to prevent contaminants from entering the system.
- **Cross-connection control:** Provide premises isolation (containment of service) or equivalent in-premises fixture protection to prevent contaminants from entering the water supplier's system.
- **Water quality monitoring:** Provide surveillance of system to detect contaminants in the water supply.
- **System operator:** Ensure that qualified personnel operate PWSs through operator certification.
- **Emergency plan:** Establish procedures for correcting problems detected in water quality monitoring or caused by natural disasters.

On the customer's premises, plumbing and health codes establish minimum design, installation, and operating requirements for public health protection. Major items in the plumbing codes are as follows:

- **Distribution:** Install approved materials and follow design requirements to ensure adequate pressure at fixtures.
- **Cross-connection control:** Provide backflow preventers at fixtures and appliances to prevent contaminants or pollutants from entering the potable water system.
- **Licensed plumber:** Require that a licensed plumber (with some exceptions, such as a landscape contractor or fire-sprinkler contractor) perform work that is plumbing code compliant.

These requirements are conservative. They include a high safety factor for system design (reliability) and for acceptable contaminant levels. For example, regulation of chemical contaminants may be based on a possible adverse health effect from the long-term (e.g., lifetime) consumption of two liters of water per day with a chemical at a level above the MCL.

Most SDWA requirements deal with possible chronic (long-term) health effects. Contamination of a water distribution system through a cross-connection may result in acute (immediate adverse) health effects that cause illness or death of one or more persons and/or financial losses. Although cross-connection control is only one of the multiple barriers to protect potable water quality, it is one of the most important. Without the water supplier's cross-connection control program, the distribution system may become the weak link in the multiple-barrier approach.

Potable water is water that does not contain objectionable pollutants or contaminants, and is considered satisfactory for drinking or culinary purposes. By this definition, potable water need not be sterile. Potable water may contain nonpathogenic organisms and other substances. For cross-connection control purposes, potable water is considered to be safe for human consumption, meaning it is free from harmful or objectionable materials, as described by the health authority. In assessing the degree of hazard, "safe for human consumption" or "free from harmful or objectionable materials" are not clearly defined parameters. A chemical toxin in high concentrations may cause no harm when consumed in low concentrations. In assessing the actual and potential degrees of hazard, microbiological, chemical, radiologic, and physical parameters must be considered. These parameters are described in the following sections.

Microbiological

Waterborne disease pathogens are the primary concern in cross-connection control. Waterborne diseases are caused by the following major groups: bacteria, virus, algae, fungi, protozoa, and parasitic. The risk to public health of a waterborne disease transmitted through the public water supply is exacerbated by the

- Large population that may be exposed to the contaminant.
- Ability to immediately detect contamination. The first indication may be a positive microbiological sample.
- Effectiveness in tracing the source. For example, *Giardia lamblia* cysts may enter the distribution system from a reservoir or through a cross-connection with an auxiliary supply.

The amount of the infectious organism ingested contributes to the difficulty of assessing the relative risk to public health from a microbiological contaminant. The health effect to an individual consuming a microbiological contaminant varies by the type of organism, the quantity ingested, and the strength of the person's immune system. For example, water with a low level of the total coliform bacteria *Citrobacter freundii* presents little adverse health concern; however, this bacterium may colonize distribution system piping and could become a health risk. By comparison, the ingestion of only a few *Giardia lamblia* cysts may be infectious.

Although a microbiological contaminant may not be a pathogen or opportunistic pathogen (one that affects a person with a weak immune system), their presence in the water distribution system may be an indirect concern. Some microbiological contaminants may cause taste and odor problems or increase the chlorine demand. A coliform bacteria detected in the water supplier's monitoring program may require mitigation measures from resampling to a boil-water order with an emergency water main flushing and disinfection program.

To assess the problem of bacteria entering the distribution system, the water supplier must consider the following issues:

- Poor-quality source water may enhance bacteria growth and regrowth in the distribution system. For example, source water with a high level of organic compounds provides a food source for bacteria that may enter the distribution system due to a backflow incident. Other quality concerns include water with high turbidity, sulfate-reducing bacteria, and iron and manganese that provide a biofilm (slime) or biomass (sediment) in water mains that facilitate bacterial regrowth.
- Distribution system piping that is in poor condition may aid bacteria growth. For example, corrosive water may cause tuberculation to form on old unlined cast-iron

and steel water mains. The tubercles provide a rough surface that shelters bacteria and a biofilm from the disinfectant.

Systems may have inadequate capacity to maintain pressure during peak water demand periods (e.g., fire flow, hot summer weather). Many old distribution systems have a relatively high frequency of breaks or leaks. Whenever there is a reduction or loss of pressure in the distribution system, there is the possibility that contaminants will flow back into the potable water system.

- Difficulty maintaining disinfectant residuals, such as chlorine, in the distribution system makes it possible for pathogens to survive.

Because each water system is different, the concerns about microbiological contamination are different for each water supplier.

Chemical

There are acute and chronic toxic effects that can occur from exposure to harmful chemicals.

The health effects of a toxic chemical vary by type of chemical and quantity ingested by the infected person. For most people, ingestion of water with a high copper level may likely cause nausea, diarrhea, abdominal pain, and/or headache. In the small portion of the population that is extremely sensitive to copper, the health effects may be poisonous, perhaps causing death.

Some chemicals have a low level of toxicity. However, when combined with the chemicals that are added to a water supply, a potentially more harmful chemical may form. Chemical contaminants may also react with the piping material in the plumbing or distribution system to leach toxic metals into the water. Because every water system treats its water differently, concerns about corrosive water are different.

Physical

There are few physical hazards that are not also chemical hazards. Examples of "pure" physical hazards include hot water and steam. Human contact with these hazards may result in burning of the skin, eyes, etc. In addition, physical hazards may also cause damage to the distribution system piping or materials.

LEGAL ASPECTS

Removing or controlling all cross-connections is a challenging task, one that could require resources beyond the financial capacity of many water systems, as well as public health and plumbing inspection departments. Frequently, property owners will modify a plumbing system, allowing uncontrolled cross-connections. Once contamination from a cross-connection occurs, it is likely that one or more persons will suffer some type of loss, e.g., a minor financial loss to cover the cost of flushing a plumbing system or serious injury or illness or death and resulting social and economic damages.

Government Statutes, Regulations, and Local Controls

Federal and state/provincial legislative bodies are heavily involved in adopting statutes that have a major impact on drinking water suppliers. Appropriate administrative agencies also promulgate regulations and periodic regulatory changes pursuant to their statutory authority. Local governments also may impose controls over water consumers through ordinances, regulations, rules, orders, and permits.

Once the statutes, local ordinances, regulations, and other administrative actions have been enacted, regulated entities are responsible for knowing and obeying these laws. Although most government agencies make an effort to notify affected parties of their newly established and ongoing obligations, contractors and builder groups should be involved in the process of enacting laws and developing regulations.

As previously stated, the primary federal statute governing the safety of PWSs in the United States is SDWA. Although major portions of SDWA are concerned with information gathering, the 1996 amendments recognized source water protection, operator training, funding for water system improvements, and public information as important new components for providing safe drinking water.

SDWA's reporting requirements may also apply to a backflow incident, whether it is the subject of enforcement or not. A variety of circumstances and events, such as failure to comply with the primary drinking water standards and other violations, must now be reported to those served by a PWS (see 42 U.S.C. 300g-3(c)). This type of required disclosure is a strong deterrent, even in the absence of civil penalties, because it exposes a water supplier to a third-party lawsuit under other statutory and common law.

Other laws and regulations that impact water suppliers and customers include

- Federal and state environmental and consumer protection regulations, including product liability laws (e.g., supply of tainted product: contaminated water)
- State requirements for the implementation of a cross-connection control program, testing of assemblies by certified testers, reporting of backflow incidents, records, etc.
- Plumbing codes and/or related industry standards (e.g., IPC, UPC, NFPA)

Water suppliers should remain aware of applicable state and local laws and regulations and consult qualified legal counsel concerning their possible application in the case of a backflow incident.

Common-Law Doctrines

Even though the water supplier has the responsibility for administration of a cross-connection control program, the building owner also plays a part in protection of the water supply within their facility. As such, all applicable building codes, fire sprinkler standards, and plumbing codes as well as any OSHA or Canadian Center for Occupational Health and Safety requirements need to be met. The building owner should also ensure that the necessary testing and maintenance of any backflow prevention equipment is conducted on their premises per requirements of the program administrator.

A common-law duty of every water supplier is to supply potable water to its customers. A water supplier's cross-connection control program should be designed to reasonably reduce the risk of contamination of the supplier's system and the water supplier's exposure to legal liability. If it is determined that a water supplier has failed to meet this duty, the water supplier could be held liable to its customers for damages proximately caused by the water supplier's breach of this duty. If other parties (contractors or other individuals) are at fault, their liability to any injured party may be determined in a similar manner, with any party found to have caused damage to another, assessed damages for some and possibly all injuries suffered.

Liability for supplying impure water has long been recognized as common law, most often for the incident of disease or poisoning that result from the violation, as well as for damage to machinery and goods suffered by commercial customers. Although case law varies from state to state, the general standard created by these cases is one of exercising reasonable or ordinary care to furnish pure water. Liability may result if this duty to exercise reasonable care and diligence in supplying water is breached. However, other cases emphasize that water suppliers are not insurers or guarantors of the quality of water supplied by them.

With respect to damages, a customer has the burden of proving all "special" or specific damages, such as reasonable and necessary medical expenses that were incurred as a consequence of the asserted breach of duty. Property damage, lost market value to property, loss of income, court costs, and any other specific items of expense may also be claimed. In addition, a customer may seek an award of general damages, which are for pain, suffering, and discomfort, both physical and emotional. There is no precise formula for computing these kinds of damages, and unless a law is in place to limit damages, they are determined by a judge or jury after considering the evidence introduced in criminal or civil litigation.

If statutory performance standards such as those included in the SDWA are violated in connection with a backflow incident, the water supplier's noncompliance with such standards will ease a claimant's burden of proof. This, in turn, will allow a claim of negligence per se for having violated the standard but will still require a showing that a customer's claimed damages were proximately caused by the instance of noncompliance. Because SDWA's record-keeping and reporting requirements generate a large quantity of data on which such actions might be based, the implementation of backflow-prevention measures becomes imperative, rather than defend against government enforcement or third-party claims of damage.

This page intentionally blank.

AWWA MANUAL

M14

Chapter **2**

Backflow Prevention Principles

BASIC HYDRAULICS

To move fluid between two points, two conditions must exist. The first condition is a pipe or conduit for a flow path. The second condition is a difference in pressure between two points. A fluid such as water tends to flow from a high-pressure source to low pressure.

Pressure is the force that is applied to a mass over an area, such as pounds per square inch (psi). In this manual, references are made to atmospheric, gauge, and absolute pressure and it is important to have an understanding of how these relate to each other.

Atmospheric Pressure. This pressure is the force exerted on an area by the atmosphere surrounding our planet. This pressure value will change with elevation or thickness of the atmosphere's layer. At sea level, the atmospheric pressure is approximately 14.7 psi. AWWA headquarters in Denver is nearly 1 mile higher than sea level and the atmospheric pressure is approximately 12.2 psi.

Gauge Pressure (psig). The pressure exerted within a piping system is typically referred to as *gauge pressure*. The zero basis for gauge pressure is atmospheric pressure. For example, an empty piping system at sea level will have a gauge pressure of 0 psi. The same system at AWWA headquarters in Denver will read 0 psi.

Absolute Pressure. *Absolute pressure* is defined as a combination of atmospheric pressure and gauge pressure. For example, if the atmospheric pressure is 14.7 psi and the pressure within a pipe is 30 psig, the absolute pressure within the pipe is 44.7 psi absolute. If the gauge on the piping system read a value of −4.7 psi, the absolute pressure would be 10 psi. It is important to recognize that absolute pressure cannot have a negative value. A pressure value of 0 psi absolute has absolutely no pressure.

A gauge value of 20 psi indicates that the pressure within the system is 20 psi greater than atmospheric pressure. At sea level, the absolute pressure within the system is 34.7 psi (14.7 psi atmospheric plus 20 psi gauge.)

A gauge value of –14.7 psi is the lowest gauge pressure possible at sea level, because the absolute pressure is zero. The water pressure measured within a distribution or plumbing system is measured as gauge pressure.

Like the pressure exerted by the thickness of atmosphere, water pressure will also increase with its elevation above the gauge. A water column 2.31 ft high will provide a pressure of 1 psi at the base. The water pressure will increase at the rate of 0.433 psi per foot of elevation. An elevated pipe 100 ft high filled with water will provide a pressure of 43.3 psi at the base.

For the purpose of this manual, atmospheric pressure will be represented by psia, and gauge pressure will be psig.

Hydraulic Grade Line. The hydraulic grade line is a mathematically derived line, plotted to represent the system pressure or head at any point in the distribution system. The elevation of the line is determined by calculating the vertical rise of water at various points in the system. To adequately maintain system pressure, it is critical that all service connection water pressure demands fall below the line under all flow conditions. If the hydraulic grade line falls below the demand, a vacuum may be created allowing a siphon to occur.

Vacuum. A vacuum occurs within a piping system when the pressure drops below atmospheric pressure or becomes negative gauge pressure. Because of this low pressure condition, water within a piping system will flow to the low-pressure point where the vacuum exists.

The shallow well's pump located aboveground creates a vacuum. Water is then pushed up the well shaft from atmospheric pressure applied at the ground water table. At sea level, atmospheric pressure can push water up a column 33.9 ft high if a vacuum of –14.7 psi exists at the top of the column. A barometric loop is a piping arrangement based on this principle is illustrated in Figure 2-1. The loop is elevated 35 ft above the highest water outlet, which will prevent any siphon or return flow from that outlet.

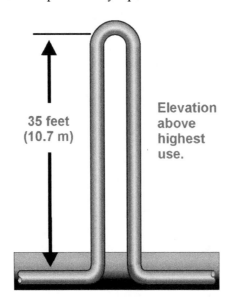

Figure 2-1 Example of barometric loop in a piping configuration
Courtesy Stuart F. Asay

Some model plumbing codes allow barometric loops and others do not. Check with the jurisdiction that has authority to determine if such a piping arrangement is an approved method for backflow prevention.

Venturi Effect. Giovanni Venturi discovered that when fluid velocity increases through a piping restriction, a jet effect is created that can produce a negative gauge pressure that may result in a siphon. A piping connection at the point of the siphon may allow backflow to occur through an uncontrolled cross-connection.

This principle is frequently used to intentionally introduce chemicals, such as chlorine, into the water line. Garden sprayers attached at the end of a hose also use this principle when applying fertilizers and pesticides.

TYPES OF BACKFLOW

In a public distribution system, the flow of water frequently changes directions to accommodate the user demand and sustain water pressure within the system. This hydraulic action is desirable and designed into the distribution grid.

Once water flows from the system through a user's water service connection, the water supplier does not want a return flow to the public distribution system. This undesirable reversal of flow is known as *backflow*.

A backflow due to a siphon is known as *backsiphonage*. Water will always flow from high pressure to low pressure. Backflow will result if the user's system pressure exceeds the distribution system pressure, which is known as *backpressure backflow*. If the pressure within the potable water piping is less than 0 psig, a vacuum is created within the pipe, and a vacuum will allow a siphon to be established.

Backsiphonage Backflow

Model plumbing codes require that a minimum residual pressure of 15 psi be maintained in the plumbing system during a flow condition. Some primacy agencies have established a distribution system design criteria for a minimum of 20 psi. This ensures the proper operation of fixtures and appliances, and it provides a positive pressure buffer to prevent the possibility of creating a siphon with negative pressure. To prevent any backflow into the public distribution system, the water supplier maintains a minimum pressure at all points in the distribution system.

Backflow from backsiphonage occurs when a subatmospheric (less than 0 psig) pressure is present in the piping system. These conditions can be illustrated by calculating the hydraulic grade line of a public water system and comparing it with all of the known water uses during all flow conditions.

Typical conditions or arrangements that may cause backsiphonage include high-demand conditions, such as a fire flow; customer demand during heat-wave emergencies; inadequate public water system source and/or storage capacity; corrosion reducing pipe capacity; water main breaks; and service interruptions.

Figure 2-2 shows how water main pressure is affected if water is withdrawn at normal and high rates. Under normal flow conditions, all service connections fall below the hydraulic grade line. Assume the hydrant at point F is opened during a period of high demand and the valve at point G has been closed, restricting the supply of water to the area or hydrant. The pressure at the hydrant drops, leaving the storage tank at point B, the top floors at point C, the house and swimming pool at point D, and the house at point E above the hydraulic grade line.

The pressure is now reduced to a point where water can no longer be supplied to these areas. To equalize the pressure, water in the lines in these areas will flow toward the lower pressures, thus creating backflow conditions.

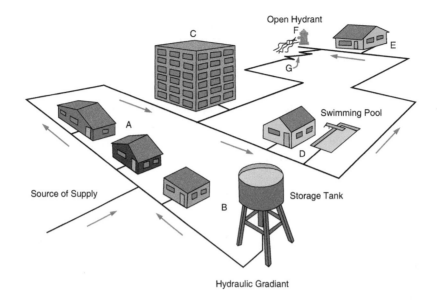
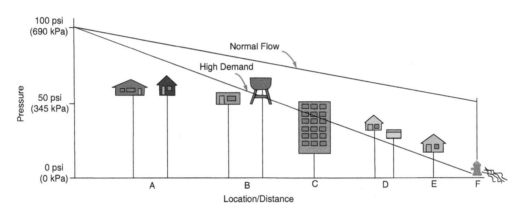

Figure 2-2   Backsiphonage backflow due to high rate of water withdrawal

The backsiphonage condition illustrated in Figure 2-3 is caused by reduced water-system pressure on the suction or inlet side of an online booster pump. The public water-main pressure is adequate to supply water only to the first and second floors; therefore, a booster pump is required to service the upper floors. The potable water supply to the dishwasher on the second floor is not protected by a backflow preventer and has a direct connection to the sewer. When periods of high demand coincide with periods of low pressure in the public water system, the booster pump that supplies the upper floors could create a backsiphonage condition by further reducing the pressure in the service connection that supplies the lower floors of the building.

The backsiphonage depicted in Figure 2-4 shows that when a distribution system is shut down to accommodate a main break repair, negative or reduced pressure will occur at all locations of the affected system that are located at any elevation higher than the break. The water main break in the street causes negative pressure in the house; this, in turn, allows contaminated water from a unprotected cross-connection to be drawn from the bathtub back toward the main. This type of backsiphonage condition usually affects more than one service connection. In fact, multiple city blocks can be affected, including commercial and industrial connections as well as residential connections.

BACKFLOW PREVENTION PRINCIPLES 15

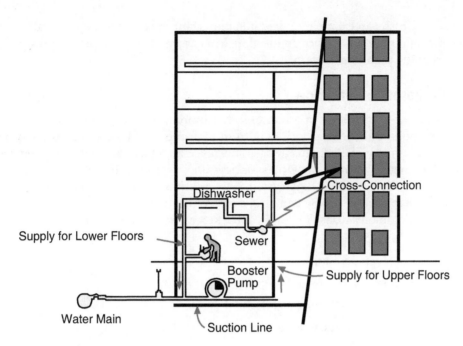

Figure 2-3 Backsiphonage backflow caused by reduced pressure on suction side of booster pump

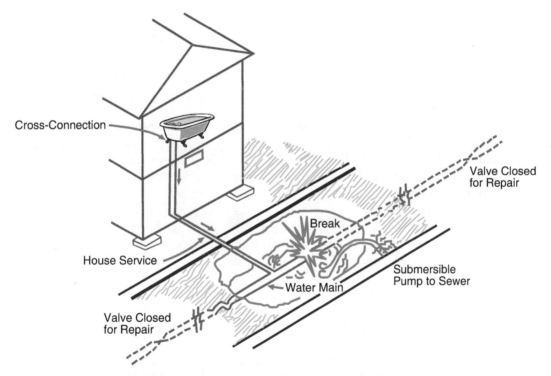

Figure 2-4 Backsiphonage backflow caused by shutdown of water system

AWWA Manual M14

This is an example of why it is important to consider the installation of a backflow preventer at each service connection. This assembly or device will prevent the backflow from the plumbing system.

Backpressure Backflow

If the water pressure in a consumer's plumbing system exceeds that of the public distribution system, water will flow from high pressure to low pressure if a backflow preventer is not installed. This undesirable reversal of flow due to the elevated pressure in the consumer's system is known as *backpressure backflow*. If an unprotected cross-connection exists within the plumbing system, the backflow occurrence may contaminate or pollute the potable water supply from the cross-connection.

Common causes or sources of backpressure include

- pumps
- elevated piping
- thermal expansion
- private wells for irrigation
- pressurized containers
- process water systems

Booster pumps are commonly used to meet fire demands or manufacturing-process demands. Pumps may also be used for chemical-feed systems, auxiliary irrigation systems, car washes, and cooling systems. If a pump is used on a property served by a public water system, there is a possibility that the pressure generated downstream of the pump may be higher than the pressure in the potable water system.

Elevated piping. Whenever a potable water system serves water to locations at elevations above the water system's source, the pressure head from the elevated column of water creates backpressure. This may produce backpressure backflow. Examples of elevated head pressure piping include water storage tanks, fire-sprinkler systems, and high-rise buildings.

Thermal expansion. Thermal expansion is a physical property related to a water volume increase inside the pipe when water is heated. All liquids and gases expand when heated. If this pressure increase is not dissipated, backpressure backflow may result. To prevent backflow, a closed piping system must have a means to safely accommodate or relieve the effects of excessive pressure caused from thermal expansion, such as an expansion tank required by model plumbing codes. For example, when raising the temperature of a 30-gal water heater from 65°F to 120°F, the water volume will expand ½ gal.

Boilers and water heaters are a common source of backpressure backflow caused by thermal expansion. Boiler makeup water must have a backflow preventer to isolate it from the potable water system in order to control a potential cross-connection.

Another source of thermal expansion is a fire-sprinkler system located at high points of buildings and subject to air temperature increases. As the temperature increases, the piping system temperature increases, the water expands, and the pressure in the piping system increases.

Pressurized containers. Gases under pressure are used in various commercial applications. One of the most widespread uses of gas is the supply of carbon dioxide for carbonated beverages. Carbon dioxide cylinders used for post-mix beverage-dispensing machines that are found throughout the food service industry contain pressures that far exceed the water system pressure. (Figure 2-5 illustrates a plumbing connection commonly used for post-mix beverage-dispensing machines.) If the safety provisions in the carbonation process fail, the water supply may be exposed to carbon dioxide gas pressures of 1,500 to 2,000 psi.

BACKFLOW PREVENTION PRINCIPLES 17

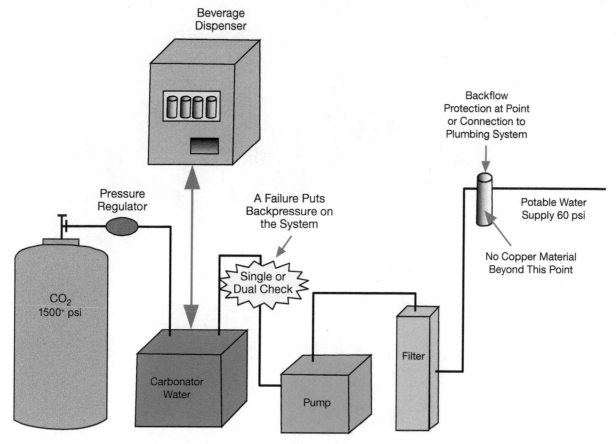

Figure 2-5   Backpressure backflow caused by carbon dioxide cylinder

Pressurized tanks are also used in many industrial facilities. For example, hydropneumatic tanks are used on pumping systems to protect motors from hydraulic shock resulting from frequent starts and stops. A hydropneumatic tank is partially filled with water or other liquid, and the rest of the tank is filled with air or another gas. The compressibility of the gas allows the liquid to be supplied within a desired pressure range without pumping.

Backflow is possible when the water supplier's supply pressure is less than the pressure of the gas cylinder or hydropneumatic tank. The volume of liquid stored in the hydropneumatic tank adds to the volume of liquid in the piping system that could flow back into the potable water system.

Figure 2-6 shows how a pump on the customer's water system can increase the water pressure to a point where it exceeds the public water distribution system pressure, causing a backflow condition.

As an example, it is a common practice to flush ships' fire-fighting systems by connecting them to dockside freshwater supplies. As shown on the graph, under normal conditions the pressure in the main is 100 psi (689.5 kPa), and it is approximately 75 psi (517.1 kPa) where it enters the ship's system.

After completing the flushing operation, a test is conducted to determine whether the fire pumps aboard the ship are operating properly. As shown on the graph, the fire-system pressure is boosted to 200 psi (1,379.0 kPa). If the valve at point A is accidentally left open, the fire-system pressure, which is higher than the public water distribution system pressure, forces salt water into the dockside and public water systems.

AWWA Manual M14

18   BACKFLOW PREVENTION AND CROSS-CONNECTION CONTROL

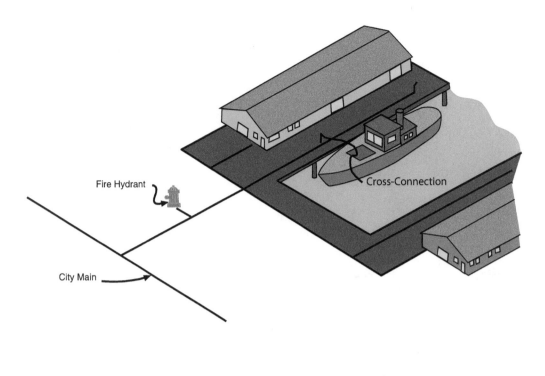

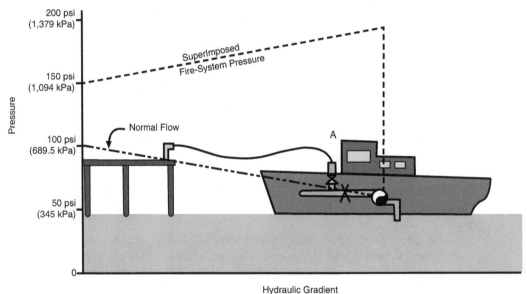

Figure 2-6   Backpressure backflow caused by pumping system

ASSESSING DEGREES OF HAZARD

The proper application of a backflow preventer depends on an accurate assessment of the risk to a potable water supply. This applies to a single uncontrolled cross-connection (i.e., a plumbing fixture) or a group of uncontrolled cross-connections (i.e., an entire plumbing system). The importance of accurate risk assessment is reflected in a common phrase used in reference manuals and regulations: "A backflow-prevention assembly commensurate with (appropriate for) the degree of hazard shall be installed …."

When an actual or potential cross-connection is identified, the essential questions to answer regarding the proper protection of the water supply are: (1) would the material entering the potable water supply from the cross-connection constitute a health or non-health hazard? and (2) would the backflow occur from backpressure, backsiphonge, or both?

Assessing Cross-Connection Contamination Risks

Every cross-connection poses a different risk based on the probability of

- the occurrence of a physical connection between a potable water supply and a nonpotable substance;
- the occurrence of backflow conditions (e.g., a water main break that causes backsiphonage);
- the failure (however unlikely) of the backflow preventer used to control the cross-connection; and
- the probability that a nonpotable substance is present and will have an adverse effect on the water system and users.

For a potable water system to become contaminated or polluted, all four factors must be simultaneously present.

Actual Connections and Potential Connections

Actual connections are common. Examples include a submerged connection of a potable water line made to a plating tank and a connection of a potable water line made to a lawn irrigation system.

The plumbing codes require that every water outlet be protected from the possibility of backflow through the use of a mechanical device or an air-gap separation. If any method is compromised, due to a failure, removal, or tampering, an actual cross-connection will exist.

Examples of potential connections are a hose-bib outlet located near a plating tank and an irrigation system supplied from a pond or other auxiliary supply located on premises that are supplied with potable water service. An outside hydrant without a vacuum breaker is a potential cross-connection. Once a hose is attached with the end submerged in a kiddie pool, an actual cross-connection exists.

One difficulty in assessing the probability of a potential connection becoming an actual connection relates to changes made to plumbing in customers' premises. For example, do-it-yourself homeowners can easily modify residential plumbing systems. Homeowners often purchase and install lawn sprinkler systems, residential solar heating systems, etc., without obtaining plumbing permits, which in some jurisdictions may not be required. Similarly, in any complex piping system (e.g., industrial plant, hospital), there is an inherent risk that maintenance personnel or other persons may modify the plumbing system removing the controls of to prevent cross-connections.

Assessing the Risk of Backflow Conditions

Backpressure conditions must be considered when assessing actual or potential cross-connection conditions. Examples of actual backpressure conditions include a connection made to a high-pressure heating boiler and a connection made to a booster pump that supplies water to a high-rise building. Examples of potential backpressure conditions include devices, such as a standpipe connection or fire-sprinkler connection, that are installed for use by the fire department to pump additional water into a fire sprinkler system, and a connection made to a hot-water tank. An increase in pressure within the plumbing system may occur if the water heater's pressure–temperature relief valve fails.

Backsiphonage is evaluated primarily as a potential condition. The following are examples of potential backsiphonage conditions:

- Water main breaks in a distribution system. The greater the number of breaks (e.g., per mile of pipe), the greater the frequency of backsiphonage conditions.
- The prevalence of elevation changes. Any water main break, fire flow, closed valve, or like condition increases the likelihood that backsiphonage conditions will occur on hilltops or tall buildings.
- Limited hydraulic capacity and redundant components in the supply and distribution system. A major reduction or loss of positive pressure may occur during high-flow conditions (e.g., peak-hour demand plus fire flow). Shutting down a major system component (e.g., a transmission line or booster pump station) may result in a major reduction or loss of positive pressure in the distribution system, increasing the likelihood of backsiphonage conditions.

The relative probability that backsiphonage conditions will occur is comparatively easy to assess. Many of the underlying causes of backsiphonage conditions are under the control or management of the water supplier and, thus, are well understood.

ASSESSING OTHER RISK FACTORS

The public health aspects of cross-connection control were discussed in chapter 1. In general, the public health risk of a nonpotable substance ranges in descending order from a health hazard to a nonhealth hazard. A health hazard may include acute microbial to acute chemical to chronic chemical contamination.

Classifying the risk of the many microbial and chemical contaminants is not practical. Therefore, the possible addition of any substance to potable water poses a risk to the water supplier's system. For example, where a nontoxic chemical is introduced to a plumbing system through a cross-connection, a risk exists that the chemical will be unintentionally replaced by a toxic chemical. Consider a post-mix beverage dispenser. Carbon dioxide gas dissolves in water and becomes carbonic acid, which dissolves copper from the pipe and the solution becomes toxic in high concentrations.

Furthermore, any water that leaves the control of the water supplier should be considered at risk of contamination. Potable water quality does not improve with age, particularly when it remains stagnant in a customer's plumbing system or storage tank.

Putting It All Together

Unfortunately, it is difficult to determine the probability of occurrence of the many events that influence the risk of contamination or pollution. For this reason, selecting a backflow preventer "commensurate with the degree of hazard" for fixture isolation or service containment remains a subjective decision. However, the lack of a backflow preventer significantly increases the potential that backflow will occur; and the greater the effectiveness of the backflow preventer, the lower the potential that backflow will occur.

From the water supplier's perspective, each of its customers' plumbing systems poses a potential risk to the potable water quality. The degree of hazard ranges from low (nonhealth hazard) to high (health hazard). In assessing the degree of hazard, the water supplier must focus on the overall hazard posed by a customer's entire plumbing system when considering a requirement for service connection protection.

ASSESSING THE EFFECTIVENESS OF ASSEMBLIES AND DEVICES

Any means designed to stop backflow, when it is working properly, can be considered a backflow preventer. These may range from a dual check valve (non-field testable device) to a field testable backflow prevention assembly that contains test cocks and shut-off valves. However, not all backflow preventers are considered equal in stopping backflow. The common types of backflow preventers are described in chapter 4.

Any mechanical apparatus can fail to perform as designed. Failure may be the result of a design flaw, operating conditions that exceed design parameters, improper installation, normal wear on moving parts, corrosion, etc. Factors that increase the reliability of a backflow preventer include

- Construction to meet design and performance standards.
- Verification of compliance through evaluation by an independent testing agency that has experience in evaluating backflow preventers.
- Inclusion of a field evaluation as part of the verification of compliance.
- Use of field-testing methods (e.g., testing a check valve in the direction of flow) that indicate substandard performance (i.e., failure to meet performance criteria) before a component actually fails.
- Regular field testing of assemblies (e.g., a minimum of once per year) to determine the need for repair or replacement.
- Assurance (through regulation) that assemblies will be maintained, repaired, or replaced if test results indicate the need to do so.
- Assurance that field testing and maintenance is performed by qualified (trained) personnel.
- Assurance that persons field testing backflow preventers are qualified through certification programs and audits of their work.
- Correct installation of backflow preventers to ensure proper operation and to facilitate testing.

The best backflow preventer is an approved air-gap separation. However, air gaps can easily be eliminated or defeated if they are not maintained in accordance with the air-gap standard maintained by the American Society of Mechanical Engineers (ASME/ANSI A.112.1.2) and the Canadian Standards Association (CSA B64). In addition, air gaps can be dangerous if they are located where they could be exposed to toxic fumes.

The reliability of backflow preventers that are designed for in-line testing and maintenance can be confirmed. Indeed, testing to verify continuing satisfactory performance is one key to an effective cross-connection control program. Generally, the assemblies are backflow preventers that require certain parts (such as test cocks and shutoff valves) that allow field testing. Assemblies must be able to be tested and repaired in line. They must meet product performance standards. Backflow prevention devices, on the other hand, are not always designed to allow field testing. The device performance standards differ and do not provide the same level of performance that reduced pressure principle assemblies provide, for example. Definitions of assemblies and devices are included in the glossary; chapter 4 discusses common types of assemblies and devices.

The effectiveness of backflow preventers can only be generally quantified. The relative effectiveness of backflow assemblies can be established by a review of field-test results. Generally, the appropriate application of and the relative effectiveness provided by backflow preventers can be assessed as listed in Table 2-1, and Table 2-2 for those in Canada.

Table 2-1 illustrates practical applications of technology. For protection of the water supplier's distribution system through a service protection policy, the choice is normally between the installation of a reduced-pressure backflow assembly and a double check-valve assembly. However, some jurisdictions consider the use of testable vacuum breakers as adequate protection for common-area greenbelt irrigation systems. Also, residential dual check valves may be installed on low hazard conditions.

Table 2-1   Means of backflow prevention in the United States

	Degree of Hazard			
	Low Hazard		High Hazard	
Means	Back-siphonage	Back-pressure	Back-siphonage	Back-pressure
Air gap (AG)	X		X	
Atmospheric vacuum breaker (AVB)	X		X	
Spill-resistant pressure-type vacuum-breaker assembly (SVB)	X		X	
Double check valve assembly (DC or DCVA)	X	X		
Pressure vacuum-breaker assembly (PVB)	X		X	
Reduced-pressure principle assembly (RP)	X	X	X	X
Reduced-pressure principle detector assembly (RPDA)	X	X	X	X
Double check valve detector check assembly (DCDA)	X	X		
Dual check device	X	X		
Dual check with atmospheric vent device (internal protection only)	X	X		

Table 2-2   Selection guide for backflow preventers in Canada

Type of Device	CSA Standard Designation	Degree of Hazard			Device under Continuous Pressure
		Minor	Moderate	Severe	
AG	—	√	√	√	No
AVB	B64.1.1	√	√	√*	No
DCAP	B64.3	√	√†	—	Yes
DCAPC	B64.3.1	√	√	—	Yes
DCVA	B64.5	√	√	—	Yes
DuC	B64.6	√	—	—	Yes
DuCV	B64.9	√	√†	—	Yes
HCDVB	B64.2.1.1	√	√†	√*	No
HCVB	B64.2	√	√†	√*	No
LFVB	B64.7	√	√†	√*	No
ASVB	B64.1.4	√	√	√	No
PVB	B64.1.2	√	√	√	Yes
RP	B64.4	√	√	√	Yes
SRPVB	B64.1.3	√	√	√	Yes

* When the recommended backflow preventer is used for this degree of hazard, zone protection with an RP backflow preventer or an air gap shall also be required.

† When the recommended device is used for this degree of hazard, zone or area protection with a DCVA backflow preventer, RP backflow preventer, or an air gap shall also be required.

NOTE: This table reflects requirements in Canada, which has moderate hazard classification as per CSA Standard B64.10-11.

"With the permission of the Canadian Standards Association (operating as CSA Group), material is reproduced from CSA Group standard B64 SERIES-11 - Backflow preventers and vacuum breakers, which is copyrighted by CSA Group, 178 Rexdale Blvd., Toronto, ON, M9W1R3. This material is not the complete and official position of CSA Group on the referenced subject, which is represented solely by the standard in its entirety. While use of the material has been authorized, CSA is not responsible for the manner in which the data is presented, nor for any interpretations thereof."

AWWA MANUAL

M14

Chapter **3**

Program Administration

The purpose of a cross-connection control and/or backflow prevention program is to avoid contamination of the public water supply by preventing, eliminating, or controlling cross-connections. The water supplier should carry out a comprehensive and effective cross-connection control program to ensure that public health is protected and the requirements of the Safe Drinking Water Act (SDWA) are complied with.

A cross-connection control program differs from a backflow prevention program in that cross-connection control provides protection in the plumbing system. Backflow prevention provides protection of the water supply distribution system and is generally installed on the service connection. The water supplier should strive for a more complete program, as cross-connection control protects the end user and backflow prevention protects the public supply. While this is an ideal approach, staffing levels or jurisdictional issues may limit the water supplier to enforcing a backflow prevention program.

Cross-connection control and/or backflow prevention programs established by a water supplier can have impacts beyond the water supplier and may require the support of many professionals including

- Plumbing inspector
- Building inspector
- Regulatory/health agencies
- Elected officials
- Public water consumers
- Fire marshal
- Backflow prevention assembly testers

DEVELOPING A CROSS-CONNECTION CONTROL PROGRAM

A well-developed, implemented, and maintained program will provide the water supplier a dependable means to accomplish the drinking water protection objectives set forth in the SDWA.

The recommended components of a program are discussed in this section.

Program Considerations

The water supplier has many choices available for effectively protecting a drinking water distribution system. Choosing a program is not a one-program-fits-all situation. The water supplier is encouraged to understand the types of cross-connection control and/or backflow programs available and implement the best option for the situation. Available programs include

- Isolating its customer's plumbing from the public supply by installing a backflow preventer on the service line. This method is known as *service protection, containment,* or *premises isolation.*
- If the backflow preventer is owned by the water supplier, it is most frequently installed on the supply side of the point of service and before the water meter.
- In most cases, the backflow preventer is owned by the water customer; however, the location of the device is determined by the water supplier. It is important to recognize that service connection protection does not necessarily protect the property occupants from internal cross-connections.
- Relying on the installation of backflow preventers at each plumbing fixture or appliance in the consumer's system. This is referred to as *internal protection, fixture protection,* or *in-premise protection.*

Mandatory service protection should be required for high-hazard categories of customers, such as

- radioactive material processing plants or nuclear reactors
- sewage treatment plants, sewage pump stations, or waste dump stations
- hospitals; medical centers; medical, dental, and veterinary clinics; and plasma centers
- mortuaries
- laboratories
- metal-plating facilities
- food-processing and beverage-bottling facilities
- car washes
- premises with an auxiliary water supply, including reclaimed water
- premises where access is restricted
- piers and docks, graving docks, boat marinas, dry docks, and pump stations
- premises with fire sprinkler systems and/or private fire hydrants
- landscape irrigation systems

The water supplier ensures that any required backflow preventer acceptable to the state, province, or local plumbing code is installed on the water service. The recommended type of backflow preventer for service protection and the reasons for the risk assessment are further outlined in chapter 6.

If a regulatory authority has not established a list of water users for mandatory service protection, the water supplier should do so. The water supplier can augment any list established by the regulatory authority unless specifically prohibited by law. The options available range from requiring a backflow preventer on all service connections to requiring a backflow preventer only on the service to a category of water consumer listed by the regulatory authority for mandatory service protection.

TYPES OF PROGRAMS

Containment, also known as Service Protection or Premise Isolation

Containment is also referred to as *service protection, containment,* or *premises isolation.* Where the water supplier accepts a service-protection type of cross-connection control program, each user must be evaluated to determine the overall health or nonhealth risk to the public water supply imposed by the water customer's plumbing system. The evaluation includes a survey (chapter 6) of each customer's potable water plumbing.

Containment may be required for any of the following reasons:

- As required by the water supplier to protect the water distribution system
- Facility determined to be a health or high hazard
- Refusal to comply with the normal steps for compliance
- Facility does not allow access to areas requiring survey
- Piping not differentiable or determined to be complex
- Piping is not readily accessible (e.g., concealed piping)
- Multiple piping systems
- Inadequate piping identification
- Facility routinely changes their plumbing configurations
- Secondary or auxiliary water sources
- Manufacture/use of industrial fluids in piping systems or facility operations
- Refusal of entry
- No current as-built/engineering drawings of the potable water system

Containment assemblies do not remove the water customer's responsibility to use appropriate backflow prevention methods.

Internal Protection

If the water supplier adopts an internal protection (fixture, zone, and area protection) program and relies on the customer's point-of-use protection, the plumbing should be surveyed to determine if the protection provided is satisfactory. The water supplier should rely on compliance with the plumbing code when assessing the overall risk to the public water supply. The water supplier may also establish additional requirements for internal plumbing protection as a condition of providing service without the installation of a backflow preventer on the service line. In either case, the customer must understand that the supplier is not "inspecting" the plumbing to determine compliance with codes or regulations or to ensure protection of the building occupants.

The water supplier may impose additional requirements for internal protection. Such requirements must be reasonable with respect to what is needed to protect the public distribution system. For example, it would not be reasonable or economically feasible for the water supplier to require all plumbing fixtures to be isolated with a backflow prevention assembly that provides the highest level of protection in lieu of placing one backflow prevention assembly on the water service.

Property owners who are also the water supplier generally implement an internal protection program, because the owner is responsible for both the public distribution and private plumbing systems. Internal protection is typically regulated by plumbing codes. This manual's focus is on water suppliers that provide water to privately owned properties. Water utilities should implement an internal protection program for all properties owned by the water supplier as the supplier is liable for both the private plumbing and distribution systems.

Comprehensive Programs

Comprehensive programs are most common when the city or town is also the water supplier and has the building code official. Their jurisdiction allows them to have a comprehensive program that combines both the containment and isolation programs into one comprehensive program. In this case, all internal plumbing isolation requirements are the primary source of protection, while containment for the protection of the water distribution system is a secondary source of protection.

This type of program provides the highest level of protection for both the customer and the water supplier. When coupled with an extensive public education and outreach initiative, optimum results and high compliance are achieved. Comprehensive surveys must be completed on a regularly scheduled basis as internal plumbing can change. New equipment can be added or tenancy and ownership can change. Program staff will need to be diligent in following up on new permits and annual testing requirements for all assemblies and devices.

For this type of program to be effective, both the building and water departments will need to support the program completely. It is an extensive initiative, but once fully implemented, the highest level of protection is now provided for all parties involved.

Joint Programs

Joint programs will be a coordinated responsibility with the water supplier, building inspection authority, fire department, and other responsible users/customers, such as a secondary distribution supply. A joint program acknowledges the specifications, standards, and codes of each industry, and develops a plan to ensure continued compliance with the appropriate jurisdiction safeguarding the potable water supply.

Many water suppliers are privately owned, owned by a city that has a service area outside of its city limits, a special water district, or owned by a city that does not have building and plumbing inspection authority.

To develop any joint program, the water supplier must

- establish each agency's jurisdiction
- enter into an agreement with the other agencies (e.g., interagency agreement between a water district and county government) defining the responsibilities of each agency
- designate the lead agency and combined program manager
- determine the procedures for all aspects of the cross-connection control program, including:
  - construction plan or permit review
  - process for inspecting/surveying of a property

- program administration, including policy for communication with property owners
- enforcement of plumbing, irrigation, and fire sprinkler codes and water supplier's cross-connection control plan requirements
- program record management
- priorities for surveying existing properties, commitment of staff time, and financial resources of each jurisdiction
- requirements for application, installation, field testing, and repair of backflow prevention assemblies

Joint programs may be operated between organizations. However, even though the water supplier may have a joint program with another agency, the supplier cannot avoid the responsibility, and thus liability arising from contaminants or pollutants that enter the public water distribution system.

Choosing a Program

Regardless of the type of program selected, it must be understood that the water supplier is responsible for any contaminant that enters the public distribution system. Reliance on the customer to prevent contamination at a plumbing system cross-connection increases the water supplier's risk or potential liability.

When establishing service policies, the water supplier should consider service protection, containment, or premises isolation. The water supplier must decide on the extent to which it will rely on the customer's internal fixture protection. This decision will be governed in part by local, state, or provincial regulations that establish mandatory service protection and allowed jurisdiction.

The benefits of services protection, premises isolation, and containment are

- Limited jurisdiction
- Ensures water of questionable quality does not enter the public water system by preventing backflow at the service
- May be appropriate for facilities with extensive potable water systems, security, or access issues
- Appropriate for facilities that may present a high risk or health hazard to the public water system

Important points regarding services protection, premises isolation, and containment:

- Does not protect quality of water inside facility's plumbing
- Could be costly
- Reduced water pressure to the facility
- Creates a closed-loop system and thermal expansion provisions will be required

The benefit of internal fixture protection and isolation is

- Ensures all points of water use are protected, thereby protecting the consumer's potable water system.

Important points regarding internal fixture protection and isolation:

- Can be difficult to manage and monitor numerous cross-connections
- Plumbing modifications may inadvertently create unprotected cross-connections
- Requires knowledgeable and experienced staff to properly identify, manage, and monitor cross-connections

- If a method is applied without containment, the risk to public water supply may increase due to number and hazard of cross-connections
- Could be costly to install multiple assemblies/devices.

The benefits of a comprehensive program are

- Complete protection from backflow into the water distribution system
- Ensures all points of water use are protected, thereby protecting the consumer's potable water system
- Ability to minimize the number of internal protection devices by planning the plumbing system into areas or zones and protecting them in groups. For example: a kitchen or a hospital or prison wing could have a testable assembly installed on the water lines that enter into that area and have a combination of testable and nontestable backflow preventers on the equipment as required. An area can be zoned off with a testable backflow prevention assembly for all nonpotable water use downstream of the assembly (areas such as process or irrigation).

Important points regarding a comprehensive program:

- The management and monitoring of numerous backflow prevention devices and assemblies will add man-power requirements. However, a properly developed backflow software program can minimize the workload and streamline the process.
- Plumbing modifications may inadvertently create unprotected cross-connections, making follow-up surveys a critical component of this type of program
- Could be costly
- Pressure reductions to the facility can be experienced; however, a proper plan for internal plumbing layouts will minimize pressure loss significantly
- Creates a closed loop system and thermal expansion provisions will be required

MANAGEMENT PROGRAMS

Commitment to Resources

Funding. The cost of the cross-connection control program should be identified in the water supplier's operating budget. To reduce and/or recover the cost of the program, the water supplier may assess the general administrative cost to all customers or to one class of customer (e.g., all commercial customers) through use of

- commodity or consumption charges ($/gallon)
- meter charges ($/month), based on size of meter
- administrative fees for annual testing
- supplemental backflow prevention assembly charge ($/month), based on size of assembly
- charge for installation permits

The water supplier may also require each customer to directly bear the survey, purchase, installation, testing, and maintenance costs of backflow prevention assemblies.

Personnel. The water supplier should ensure that there are adequate personnel and resources to conduct the necessary field and administrative requirements for a cross-connection control program. The water supplier may want to incorporate the use of the AWWA M14 Manual as a guide to prevent, eliminate, and control cross-connections.

One person should be selected as program administrator/coordinator. The various tasks in operating the program may be performed by different staff members, consultants, or contractors. However, to ensure consistency and accountability, one person must manage and provide direction.

Cross-Connection Control Plan

The first action in implementing a cross-connection control program is for the water supplier's management group to formally establish a cross-connection control plan. This plan involves making a commitment to employing staff, establishing policies that involve risk and liability decisions, setting program goals, etc. A cross-connection control plan may be established by regulation, tariffs, or ordinance/by-law and can also be established by incorporation into an adopted water system plan.

Stakeholder meetings. A stakeholder meeting should be conducted to identify and discuss the cross-connection control program objectives. Identifying and mitigating coordination issues in advance through effective communication may help to improve the overall effectiveness and compliance of the program. The education of the department staff is essential in order for the program to operate correctly and effectively. All staff involved with the program should be trained and possess a thorough understanding of cross-connections. The training and experience component may include plumbing or building experience, have held a similar position (Cross-Connection Control Surveyor) with another water supplier, or a minimum of several years' experience in conducting cross-connection control surveys in residential homes and commercial, institutional, and industrial facilities.

Public outreach and education program/customer communications. Success in operating a cross-connection control program is often linked to the public's understanding of the public health risks posed by cross-connections. No one would connect plumbing fixtures, equipment, etc., to their drinking water system knowing it could pollute and/or contaminate their drinking water.

A public education program is an integral part of a cross-connection control program. Public education programs vary based on the water supplier's resources and the types of customers served. For all education efforts, the following information should be conveyed:

- The nature of the public health risk posed by actual or potential cross-connection hazards
- The fact that the water supplier is responsible for protecting the public water system from contamination or pollution
- The fact that the customer is responsible for preventing a contaminant or pollutant from entering their plumbing system and thereafter entering the public water system
- The fact that the water supplier is required to comply with state or provincial regulations concerning cross-connection control
- The fact that the water supplier has established policies (conditions of service) relating to cross-connection control

Audience. Public education efforts should be tailored to the intended audience. Following are examples for two major categories.

Single-family residential customers. For single-family residential customers (i.e., the general public), educational materials should explain what constitutes a cross-connection, how backflow can occur, etc., in easy-to-understand, nontechnical terms. The material should be kept to a reasonable length and use simple graphics to illustrate key points and keep the reader interested.

Common forms of public education for this group of customers include

- water bill inserts (brochures)
- consumer confidence reports
- information on the service policy distributed with application for service for new customers
- mall, fair, home show displays, and similar venues
- use of the news media

The annual consumer confidence report (CCR), required by the USEPA, should contain a brief statement describing the water supplier's operation of a cross-connection control program and the customer's responsibility to protect their plumbing system and the water supplier's water distribution system. Educational brochures should be distributed or made available on the Internet or the supplier's web site.

Public service announcements on radio and television are very effective. The radio announcements are generally free, but there are costs for developing a short video(s) for television. Newspaper articles alerting the consumer to the dangers of unprotected lawn sprinkler systems, using a hose to spray weed killer and fertilizers, or unplugging drains are examples of article topics.

Commercial customers. This group includes all customers other than the single-family residential group. The educational information provided to commercial customers may be more technical. It should be directed to the specific customer (e.g., dry cleaner, dental clinic, shopping mall owner, restaurant, etc.). Commercial customers are likely to confer with their maintenance staff, plumber, architect, etc., regarding the information provided. In many cases, the cost to the commercial customer of installing backflow preventers on existing premises is significant compared to their operating budget. This may generate complaints or other resistance regarding compliance that must be addressed with educational materials.

Professional and trade groups. When providing education to technical groups, the supplier should stress that customers rely on their consultants, contractors, etc., to fulfill the customer's responsibility to prevent contamination. Examples of technical groups include

- Occupational Safety and Health Administration inspectors,
- health inspectors,
- architects and engineers,
- building and/or plumbing inspectors,
- plumbing suppliers,
- plumbing contractors and plumbers,
- wastewater treatment personnel,
- irrigation contractors and suppliers,
- fire suppression contractors,
- pool and spa contractors,
- pest control product suppliers, and
- pipe fitters and mechanical contractors.

Smaller water suppliers located near large utilities with good cross-connection control programs can join forces. In some areas, this can be done through local AWWA sections or by establishing a local cross-connection control committee. These efforts promote good public relations. Persuasion with understanding is preferable to enforcement of regulations. Aware customers are more likely to maintain their backflow preventers and are more tolerant of the water supplier for imposing the added costs of backflow prevention.

Program Performance Goals

Management should periodically review the program's implementation. Periodic changes may be necessary in cross-connection control plans, policies, and procedures if the program goals are not being met or if the program needs to be expanded or amended. Corrective action may include customer correspondence, public education, on-site visits, administrative fees, fines, termination of service, work orders, and legal action.

Coordination with Local Authorities

Coordination with other authorities involved in the cross-connection control process may range from exchanging program information to operating a joint program. As a minimum, the water supplier should inform local regulatory agencies (e.g., building, plumbing, and health officials) of the following:

- The water supplier requirements (service policies), e.g.,
  - ensuring that all premises shall have a backflow preventer. For the purpose of ensuring that thermal expansion is compensated for, the plumbing inspector will need to enforce the plumbing code requirement to compensate for thermal expansion for protection and a pressure-temperature relief valve on the water heater or
  - ensuring that all backflow prevention assemblies installed at the service for protection or at the internal fixture in lieu of service protection must be on the water supplier list of approved backflow preventers
- The results of the water supplier's survey of premises, including
  - the notification given to a customer of the specific requirement for service protection or internal fixture protection in lieu of service protection
  - violation of the water supplier's cross-connection control requirements (e.g., for failing to submit an annual field test report)
  - the notification given to a customer that water service will be discontinued for cause.
- The receipt of water quality complaints that may indicate a cross-connection control problem.

These coordination efforts by the water supplier are made primarily for risk management. They are also beneficial for public education. Therefore, these efforts should be made unilaterally.

In return, the water supplier should request the following from the local authorities:

- certificate of occupancy
- that the local authority notify permit applicants that the water supplier may have separate (additional) cross-connection control requirements
- a copy of the results of their plan review and/or inspection of a plumbing system with regard to cross-connection control (e.g., any health hazard requiring a reduced-pressure principle backflow prevention assembly)
- immediate notice of any enforcement action taken with regard to cross-connection control

- immediate notice of any water quality complaint (even if it is not investigated by the local authority)
- immediate notice of the investigation of a water quality complaint that involves cross-connection control

As part of the water-supplier risk-management program, the request for this coordination effort by a local authority should be made in writing to the management of that authority.

Enforcement. Historically, any primary enforcement action conducted by water suppliers has been to shut off or discontinue water service. Although immediately effective in alleviating the risk of backflow, the termination of water service can create a health hazard, property damage, or substantial financial loss for the customer.

Although the water supplier may have the right and does have the obligation to terminate service to prevent backflow, the exercise of this right should be judicious. A customer's failure to submit a field-test report within 30 days does not constitute an immediate risk to the water supply. It should be noted that service interruption may be required by local codes, rules, or ordinance. Any enforcement action by the water supplier made on the grounds of a "public health risk" may be challenged by the customer in a court of law. If the actions of the water supplier are found to not be supported in the governing code, rule, or ordinance, the water supplier may have to pay substantial compensatory and punitive damages.

For these reasons, the water supplier should establish well-defined guidelines for enforcement actions, which may include administrative fees, fines, work orders, or installation by the water supplier of a service protection backflow preventer on the service pipe rather than discontinuing water service. The enforcement policy should address the following:

- State and provincial, local laws and/or regulations
- The effort to notify the customer of the requirement before the enforcement action is taken. (For example, mail first notice, mail second notice 30–45 days thereafter, mail final notice 7–10 days after the second notice. One [or more] notice[s] should be sent via certified mail, return receipt requested.)
- The appeal procedures of any enforcement action
- The criteria for immediate shutoff of water (e.g., impending hazard, meter running backward, backflow occurring)
- Notification of customer noncompliance to other regulatory authorities (e.g., state/provincial and local health departments, building inspectors.)
- Variation in policies for different water services (e.g., fire lines) or category of customers (e.g., apartment buildings)

The water supplier should consult its legal counsel when instituting an enforcement policy and prior to taking any enforcement action that could cause a major financial loss or operating problem for a customer.

When establishing an enforcement policy, the water supplier should consider the advantages of a written service agreement with the customer. The customer's failure to comply with the water supplier's requirements for the installation and maintenance of backflow preventers would constitute a breach of condition of service by the customer. Remedial action (e.g., shutoff of domestic water service) would be specified in the contract.

DOCUMENTATION

Record Keeping

Good record keeping is essential to the proper operation of a cross-connection control program. Additionally, records may be subject to audit by state or provincial regulatory agencies. All original records (correspondence, plans, etc.) should be kept in the water supplier's files. If contractors are utilized (consultant cross-connection control specialists), the water supplier should retain copies of all records (paper and/or electronic). Proprietary, industry-recognized database software programs developed specifically for managing a cross-control program are commercially available.

Record of risk assessment. For each customer, the water supplier shall have a record of the initial risk assessment and subsequent reassessment, in the form of a completed water-use questionnaire (all water customers) and a cross-connection survey report.

For risk management reasons, the water supplier should retain both the initial form plus the latest reassessment form, because both could

- demonstrate that the water supplier has complied with the state/provincial requirement to evaluate new and existing customers to assess the degree of hazard
- contain a signed statement from the customer or customer's contractor (e.g., backflow prevention assembly tester or cross-connection control specialist) about their water use and/or assessed degree of hazard
- contain information useful for the investigation of a backflow incident

Control of Records

The water supplier should maintain all correspondence with its customers for a minimum of 7–10 years or as required by local jurisdiction and state or provincial authority. The most current service agreement and instructions to install backflow preventers to protect the water supplier's water distribution system should be maintained as a permanent record. All correspondence with the state or provincial authority and the local administrative authority should be maintained for at least 7–10 years.

HUMAN RESOURCES

Training

Proper field testing of backflow preventers and assessment of hazards requires specialized training and/or experience. Certified backflow prevention assembly testers and certified cross-connection control program specialists and administrators can be trained at various locations throughout North America. Training should include but not be limited to seminars, workshops, conferences, courses that may be provided by local and state jurisdictions, provinces, industry-recognized technical organizations (e.g., AWWA sections), colleges, manufacturers, code agencies, and consultants.

Training is available for

- cross-connection control program administration
- backflow prevention assembly field testing
- backflow prevention assembly repair
- cross-connection control surveys

Not all training courses are equivalent. Training courses within North America, a geographical region, or even a state or province may vary in the technical content, instructor's qualifications, course length, etc. For example, training courses may vary in scope, ranging from an introduction to cross-connection control to basic training to continuing education to advanced training. (Benchmark training will, at a minimum, total approximately 32–40 hours.) Some training courses may be directed to specific groups (e.g., plumbing inspectors).

In selecting training for its staff, the water supplier should consider the following:

- Where state or provincial courses are available, determine whether the content is satisfactory, and whether supplemental training is appropriate to improve quality assurance.

- Where state or provincial courses are not available, evaluate and select the other sources of training, taking into consideration such things as the official recognition of the training course by the state or province; the scope and content of the subject matter; the instructors' qualifications; and the quality of the training materials, facilities, and equipment.

- Provide supplemental continuing education through seminars, workshops, refresher courses, and conferences on cross-connection control.

- Provide publications relating to cross-connection control as an important source of supplemental continuing education.

Consider participating in regional cross-connection control technical groups to exchange current information (e.g., an AWWA section's cross-connection control committee).

Certification and Competencies

Cross-connection control backflow prevention assembly testers should be certified to ensure the minimum level of proficiency needed to perform the task of field testing backflow prevention assemblies. A certification program may be administered by a state or province, an industry recognized technical organization, local administrative authority, or the water supplier. Where the state or province requires certification, the water supplier may require more stringent requirements to a higher level of proficiency.

As a minimum, the water supplier should require that certification for backflow prevention assembly testers (staff and/or contractors submitting field-test reports) include

- successful completion of a certification exam acceptable to the administrative authority (such exams incorporate both written questions and hands-on testing of backflow prevention assemblies overseen by a competent proctor independent from the training course)

- periodic recertification as required by the certifying authority through a written and hands-on exam. Recertification should at a minimum require demonstration of competency through a written and hands-on performance examination, which may also include completion of continuing education.

- revocation of certification for backflow prevention tester misconduct or improper testing of assemblies

Safety

Operation of a cross-connection control program may involve the water supplier's staff who will install, test, and maintain backflow preventers and survey customers' premises for cross-connections. Consequently, the water supplier's safety program should recognize the following issues, which relate specifically to cross-connection control:

- Repair of a backflow prevention assembly requires special tools and specialized education/knowledge. Use of the wrong tool may result in injuries to the worker or damage to the backflow prevention assembly. For example, metal tools used to clean seats may damage them, and large assemblies may have heavy covers, and their springs may be under great tension. Remember, only manufacturer's parts may be used for repair.
- Many backflow prevention assemblies are installed in hazardous locations, particularly old backflow prevention assemblies installed on the customer's premises. The potential hazards related to these assemblies include confined spaces, near ceilings requiring platforms, exposure to hazardous or toxic materials in industrial plants, and automobile traffic or moving equipment.

PROGRAM ADMINISTRATION

Determining Risk

At locations where a service connection backflow preventer is not installed, the water supplier has the responsibility to periodically reassess the risk. For example, the plumbing may be changed by the owner without a permit, resulting in the creation of a hazard. Once initial surveys are complete, then a re-survey frequency should be determined for each water customer based on the degree of hazard and backflow potential. In general, situations where backflow could cause illness or death shall be considered a high or health hazard, water customers that pose a high hazard or have a high potential for backflow to occur, must be reassessed frequently. Other factors such as new construction, water quality complaints may prompt an immediate re-survey.

The backflow preventer must be appropriate with the evaluated degree of hazard. It must be in a safe location, away from toxic fumes, and should be protected from the elements. If the maximum protection is not provided, a change in the customer's plumbing may require a change in the type of backflow preventer relied on for service protection.

Where the water supplier relies on the customer's internal fixture or appliance protection, the effort to assess and reassess the risk to the water supplier is much more involved, and thus time consuming. This is an important consideration when establishing a policy to rely on internal plumbing system protection.

Vulnerability Assessment

Before water service is provided to a customer, the water supplier should perform an assessment of the system and establish any requirements for backflow prevention. If occupancy is unknown, the highest degree of protection should be required and a survey should be conducted after occupancy. The methods of assessing risk and selecting a backflow preventer commensurate with the degree of hazard are discussed in chapter 6.

Although new water services can be easily surveyed for the degree of hazard, this provision does not address established customers receiving water before the cross-connection control program was implemented. To ensure that all hazards are identified and properly controlled or eliminated, a program to survey existing water users is recommended. This program should include

- assessing any actual or potential risk from existing customers,
- notifying customers of the assessed risk and required backflow preventer(s),

- ensuring that backflow preventers are installed on existing services or at internal plumbing fixtures in lieu of service protection, and
- periodically reassessing the risk

The water supplier must establish a schedule for accomplishing these provisions. A priority for implementation must be assigned based on degree of hazard. For risk and liability management, the water supplier should prioritize as follows:

- Customers requiring mandatory service protection
- Commercial customers, industrial customers, institutional customers
- Fire protection customers
- Multifamily residential customers
- Single-family residential customers

The priority list may be further refined with subcategories as needed for scheduling work.

Management should establish the schedule for completing these tasks. Until the cross-connection control program is fully established, the water distribution system is at risk. The decision to take a risk and incur potential liability should only be made by the water supplier's management, including any elected officials.

Backflow Prevention Assemblies—Ownership

There are two possible scenarios for the installation and ownership of the backflow prevention device.

The water supplier may install the service protection backflow preventers along with its water meter. The main advantages to having the water supplier own the backflow prevention assemblies are

- assurance that the backflow prevention assemblies will be properly field-tested and maintained
- assurance that the backflow prevention assemblies are appropriate for the application
- the cost of installation, testing, and maintenance may take advantage of the economies of scale (e.g., purchase of a large number of units could provide a volume discount)
- the cost of the backflow prevention assembly may allow for an increase in rates (for investor-owned utilities) to recover program-related expenses.
- less need to enforce the water supplier's policies for the customer's failure to field-test and maintain the backflow prevention assemblies

The main advantages to the water supplier of the water customer installing the backflow prevention assemblies are

- all costs will be directly borne by the customer;
- the customer will have the responsibility, and the liability, for the proper installation, field testing, and maintenance of the backflow prevention assembly
- the backflow prevention assembly may be installed in the customer's building where protection is provided from freezing and vandalism
- the customer can select the backflow prevention assembly manufacturer that is best for their water service conditions (e.g., pressure loss, size, orientations, etc.)

Testing

Minimum criteria for acceptance of assemblies and field test equipment. To verify that a backflow preventer functions properly, it should be field-tested upon installation, after any repair, upon relocation at least annually, as discussed later in this chapter. The water supplier is responsible for ensuring that the backflow preventers are field tested by staff or a person who has demonstrated competence to test, repair, and maintain backflow prevention assemblies as evidenced by certification that is recognized by the water supplier. The water supplier may prescribe what training/certification is required.

Testing of backflow prevention assemblies. Most backflow prevention assemblies designed for field testing can be tested in place. The water supplier should field test or require to be field tested the backflow prevention assemblies it relies on under these circumstances, and to ensure proper working condition.

The water supplier should adopt installation and testing standards through one of the plumbing codes as a part of its service policies or make reference to specific industry or association recognized standards and specifications that incorporate both testing and installation standards.

The water supplier must specify the minimum standards/criteria for the backflow prevention assemblies approved and relied on to protect the public water distribution system. This may be done by establishing an approval list that includes specific manufacturer's models for each type and size of backflow preventer, installation requirements for each listed backflow preventer, and accurate field-test equipment and calibration records. In addition, an approval list established by another authority or organization can be adopted.

State or provincial governments often publish or provide a list of approved backflow preventers or specify the approval list of an organization. Approval lists exclude products. Where the approval list used by the water supplier is mandated by a government agency, that agency must show reasonable grounds for inclusion or exclusion of a product from the list. Where the water supplier voluntarily adopts a list or refines a government agency's list to exclude a product or manufacturer, the water supplier may be called on to explain the reasonableness of its action. A water supplier has the latitude to adopt additional product performance requirements to develop their own approval list.

A list of approved backflow preventers should be based on, but not limited to, the following:

- Manufacturer's compliance with nationally recognized standards (e.g., ANSI/AWWA C510 and C511, Double Check Valve Backflow Prevention Assembly, Reduced-Pressure Principle Backflow Prevention Assembly, respectively and/or others as described in the jurisdiction)
- Testing by an approved laboratory to show that the backflow preventer complies with an industry recognized standard (the testing should include a field evaluation program)
- Inclusion of conditions for backflow preventer installation where appropriate to performance (e.g., horizontal installation only)
- Periodic review for renewal of listing

Where a state or provincial approval list is established, the water supplier may refine the list by establishing more stringent requirements. When doing so, the water supplier must set clear guidelines for inclusion or exclusion from its approval list. Review by legal counsel is recommended.

To ensure the quality of the field testing of backflow preventers, the water supplier should also specify minimum standards for the field-test equipment. This may be done by establishing an approval list or by adopting an approval list established by an industry recognized authority or organization. The state or provincial government may publish a list of approved field-test equipment or specify the approval list of an industry recognized organization. All field-test equipment should be checked for calibration at least annually.

Inventory of Backflow Prevention Assemblies

An inventory of backflow prevention assemblies or air gaps, if required by the water supplier to protect its water distribution system and installed at facilities owned by the water supplier, should include

- location of the backflow preventer or air gap (adequate details to locate the backflow preventers)
- description of hazard isolated
- type, size, make, model, static line pressure, and serial number of backflow prevention assembly or air-gap details

Backflow Prevention Assembly Test Reports

An inventory for each field test or air-gap inspection should include at a minimum, the following:

- Name and certification number of the backflow prevention assembly tester
- Field-test results
- Repairs performed to obtain acceptable test results
- Repair history
- Tester's signature
- Type, size, make, model, and serial number of backflow prevention assembly or air-gap details
- Positive identification of the backflow prevention assembly test, which may include field tagging
- Test equipment calibration information
- Date and time of test

Quality Assurance Program

A quality assurance program is important for all aspects of operating a water supplier. For cross-connection control, the minimum quality assurance program should include the following:

- Review of the performance of certified backflow prevention assembly testers, which includes:
  - spot-checking (auditing) the tester's work by observing the backflow prevention assembly field test or by inspecting and retesting assemblies
  - comparing field-test data with manufacturer's data and previous field-test reports
  - checking on the proper completion of field-test report forms
  - verifying with the certification agency that the tester's certification is valid
- Review of the field-test results submitted for backflow preventers to determine
  - if the results are unsatisfactory (i.e., component[s] failed performance criteria)
  - that replacement or repair is needed

- that the backflow preventer has been replaced, relocated, repaired, modified, or removed without the water supplier's prior knowledge
- that the backflow preventer is improperly installed or in an improper application
• Monitoring of field-test equipment to ensure that accuracy is within tolerances. This may be done by
  - requiring that field-test equipment have its calibration checked by an independent manufacturer certified or industry recognized laboratory
  - having the water supplier check verification of the field-test equipment annually, at a minimum

Backflow Incident Reporting/Backflow Incident Response Plan

Details on the investigation and subsequent corrective action taken for reported backflow incidents should be kept on file indefinitely. Some state and provincial authorities require that backflow incident reports and details of incidents be submitted to them. Regulatory agencies may have a required backflow incident form that must be completed. These are sometimes required by the state primacy agency to be included in the CCRs.

Administration of a cross-connection control program should include the investigation of any water quality complaint that indicates possible backflow contamination. Quick response by personnel trained in cross-connection control and basic water quality is necessary for the success of any investigation of a backflow incident and prevention of further contamination. The water supplier should act in cooperation with other authorities or agencies to protect the consumers of water within the premises where a contaminant or pollutant has been detected.

There are documented cases of cross-connections causing contamination or pollution of drinking water. An overview of the potential for contamination and pollution, the difficulties in identifying the source of contamination or pollution, and the efforts to remove a contaminant or pollutant from a public water system can be obtained by reviewing published backflow incidents (see appendix D).

The water supplier's investigative actions are intended to

- protect the water supplier's water distribution system from the spread of a contaminant or pollutant detected in the water supply on private property
- quickly restore the quality of water in the supplier's distribution system if a contaminant or a pollutant has entered the system through a backflow incident from the customer's plumbing system
- prevent any further contamination of the supplier's water distribution system
- maintain consumer confidence in the system

Written guidelines on how to respond to all foreseeable complaints should be included in the water supplier's cross-connection control plan and operations manual. Even when written guidelines are provided, it is impossible to cover all types of water quality complaints and other operating scenarios within the water supplier's system that may contribute to a water quality complaint.

Most backflow incidents such as the backflow of "used" water, water with a low level of chemical contamination, or water with an undetectable bacteriological contaminant, are not likely to be identified as such. A person may complain to others, but not to the water supplier, about water with a slight taste or odor. When a complaint is made to a water supplier, the water supplier may not respond expeditiously and thus may not find conclusive evidence that a contaminant has entered the potable water system through a cross-connection.

When the initial evaluation of a water quality complaint indicates that a backflow incident has occurred (potable water supply has been contaminated/polluted) or may have occurred, or when the reason for the complaint cannot be explained as a "normal" aesthetic problem, a backflow incident investigation should immediately be initiated. It is wise to be conservative when dealing with public health matters.

A backflow incident investigation is often a team effort. The investigation should be made or (initially) led by the water supplier's cross-connection control program administrator. If the administrator does not have training in basic water quality monitoring, a water quality technician should be part of the investigation team. One or more water distribution system operators may be called on to assist in the investigation and to take corrective action. The investigation team may also include local health and plumbing/building inspectors.

The water supplier is involved in any backflow incident investigation because they received the complaint and have an obligation to make a reasonable response and because they have an interest in preventing the spread of a contaminant or pollutant through their water distribution system.

Each investigation is unique and requires the development of a specific plan of action at the start of the investigation, with modifications to the plan as the investigation progresses.

The water supplier's investigation should include these steps:

- Locate the source of the contamination.

- Isolate that source to protect the water distribution system from further contamination.

- Determine the extent of the spread of contamination through the distribution system and provide timely, appropriate notification to the public and to regulatory agencies.

- Take corrective action to clean the contamination from the distribution system.

- Restore service to the customers.

The traditional solution to distribution system contamination, flushing by the water supplier, must be thoroughly evaluated before implementation because flushing can, in some cases, contaminate more of the system.

The public health authority must consider the needs of those who may have consumed or used the contaminated water. The customer must take any action necessary to clean the plumbing system, shut down equipment, vacate the premises, or similar steps.

In any circumstance where a claim for damages may follow, the water supplier should establish (with the advice of their attorney or risk manager) the procedures to be followed to provide advice or assistance to the customer. Additionally, the water supplier should complete a backflow incident investigation report.

The water supplier may need to notify the public of contamination of the distribution system. A public notification procedure should be part of the water supplier's general emergency plan. Failure to provide timely, appropriate notification to all customers that may be affected by water system contamination may result in a claim for damages by customers. Appropriate notification procedures are essential for liability management.

The follow-up to a backflow incident investigation should include debriefing of personnel, reviewing response and investigation procedures, and accounting for the cost of the investigation.

AWWA MANUAL

M14

Chapter **4**

Backflow Prevention Assembly Application, Installation, and Maintenance

A backflow preventer is an assembly, device, or method that prohibits the reversal of flow of liquids into a potable water supply system, which may occur through a cross-connection. The type of backflow (backpressure and/or backsiphonage) must be identified, and the degree of hazard (high or low) must be determined. In addition, installation criteria, hydraulic conditions, and the location of shut-off valves must be evaluated before selecting a backflow preventer. All of these criteria must be evaluated by trained and/or certified personnel before a backflow preventer can be installed.

Manufacturers' products may vary in that the products may prevent backflow in different ways. To ensure that products offer an acceptable level of protection, several independent organizations have developed design and performance standards for backflow preventers. These standards can vary in their requirements that a backflow preventer must meet. In addition, the local administrative authority should review the various standards and select one that provides continual and proper protection.

There are two categories of mechanical backflow preventers: assemblies and devices. Assemblies are mechanical devices that are in-line performance testable and repairable with two properly located approved shutoff valves and properly located test cocks. They prohibit the backflow of nonpotable water into the potable supply system through a cross-connection and must meet an approval standard for performance and design. Devices are not recognized as performance testable; however, they prohibit the backflow of nonpotable water into potable water supply systems through a cross-connection. Standards differ for types of devices and assemblies. Various standards describe different performance requirements based on the level of protection required.

After a backflow preventer is selected, it must continue to work as designed. To ensure this, a field-testing and repair protocol must be followed. Backflow prevention assemblies must be periodically tested after installation to verify that they continue to prevent backflow properly. In addition, the backflow prevention installation should be evaluated for modifications, changes in plumbing, and appropriateness of assembly for hazard type, and verification that the backflow preventer has not become submerged. Nontestable devices require periodic replacement to ensure proper operation. Replacement frequency will depend on local jurisdiction.

MEANS OF PREVENTING BACKFLOW

The following sections describe applications of technology intended to prevent backflow. The functional capabilities of the installations or equipment and the recommended level of protection and hydraulic limitation applications are also described.

BACKFLOW PREVENTION DEVICES

Backflow prevention devices are not to be substituted for applications that require backflow assemblies because they usually do not include shutoff valves or test cocks, and usually cannot be tested or repaired in-line. Many devices have restrictive head loss and flow restrictions. These devices are used for internal protection and come only in size 2 in. [51 mm] and smaller. The application of these devices falls under the jurisdiction of the plumbing code and is located in private plumbing systems. A water supplier that has concerns about the use of devices should consult with local plumbing code officials. Common symbols used for backflow prevention devices are illustrated in Figure 4-1.

Dual Check

Description. A dual check shall contain two internally loaded, independently operating check valves. A dual check shall be a device approved by an approval agency acceptable to the local administrative authority.

Function. A dual check contains two loaded checks (Figure 4-2). In a backpressure condition, the increase in pressure will force the checks to close tighter. If the second check is not working, the first check can act as a backup to stop the backpressure from going through the device. In a backsiphonage condition, a subatmospheric condition is present at the inlet, and the loading of the checks will cause the checks to close. Foreign debris or deterioration of the check components can affect both checks simultaneously, rendering the dual check incapable of preventing backflow without an outward indication of failure.

Application. A dual check can be used to stop backflow from backpressure and/or backsiphonage and should be used only for low hazard internal protection applications. A dual check does not provide the same level of protection as a backflow prevention assembly and should not be used for service line protection unless it is a low hazard situation.

Installation criteria. A dual check must be installed in the orientation as it was approved by the approval agency recognized by the jurisdictional authority and as allowed by regulatory authority. A dual check should be sized hydraulically, taking into account both volume requirements and pressure loss through the device.

A dual check must not be subjected to conditions that would exceed its maximum working water pressure and temperature rating. The increased pressure occurring from a closed system must also be evaluated, because excessive pressure can damage the device or other plumbing components.

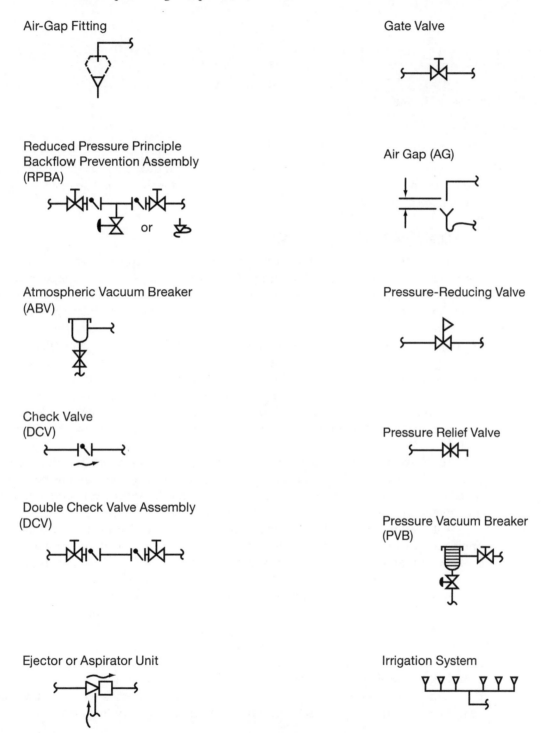

Figure 4-1   Common symbols used for backflow prevention devices

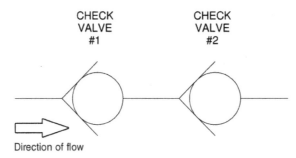

Figure 4-2 Dual check device

Courtesy of Apollo Valves/Conbraco Industries, Inc.

A pipeline should be thoroughly flushed before a dual check is installed to ensure that no dirt or debris is delivered into the device. A dual check shall be installed where it can be inspected or replaced as necessary.

Dual Check with Atmospheric Vent

Description. A dual check with atmospheric vent shall contain two internally loaded check valves and a vent valve. The vent valve, located between the two check valves, shall open when subjected to backpressure. A dual check with atmospheric vent shall be a device approved by an approval agency acceptable to the local administrative authority.

Function. The device shall have an inlet check valve that opens as flow begins (Figure 4-3). After the water has passed the first check, it shall cause the vent valve to close, allowing water to travel past the second check valve. In a backpressure condition, the increased pressure at the outlet will cause the second check valve to close. If the check does not close, the increased pressure will travel across the second check and cause the vent valve to open as it is subjected to backpressure. In a backsiphonage condition, the inlet pressure will be reduced to a subatmospheric pressure and cause the check valves to close.

Application. A dual check with atmospheric vent can prevent backflow caused by backpressure and/or backsiphonage. The device should be used only for low hazard applications and for internal protection. A dual check with atmospheric vent does not meet the requirements of a backflow prevention assembly.

Installation criteria. A dual check with atmospheric vent must be installed in the orientation as approved by the agency recognized by the jurisdictional authority. A dual check with atmospheric vent must not be subjected to conditions that would exceed its maximum working water pressure and temperature rating. The increased pressure that can happen from the creation of a closed system also must be evaluated.

A dual check with atmospheric vent should be sized hydraulically, taking into account both volume requirements and pressure loss through the device. A pipeline should be thoroughly flushed before a dual check with atmospheric vent is installed to ensure that no dirt or debris is delivered into the device. A dual check with atmospheric vent shall be installed where it can be inspected or replaced as necessary.

A dual check with atmospheric vent* may discharge water from its vent. Care should be taken to ensure that any discharge will not harm the surrounding area.

---

* Atmospheric vents should not be subject to submersion or toxic or corrosive fumes.

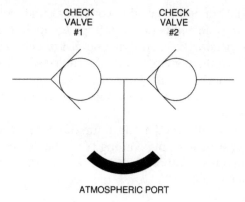

Figure 4-3   Dual check device with atmospheric port
Courtesy of Apollo Valves/Conbraco Industries, Inc.

Atmospheric Vacuum Breaker

Description. An atmospheric vacuum breaker (AVB) shall contain an air-inlet valve and a check seat. The device shall be approved by an approval agency acceptable to the local administrative authority.

Function. Water will enter the inlet of the AVB and cause the air-inlet poppet to seal against the air-inlet seat (Figure 4-4). After the poppet is sealed, water will flow through the AVB into the piping system. In a backsiphonage situation, the inlet pressure will be reduced to a subatmospheric pressure, causing the poppet to fall off the air-inlet seat and rest on the check seat. Air will enter the air-inlet port(s) to break any vacuum.

Application. An AVB shall be installed to prevent backflow from backsiphonage only. An AVB can protect both high and low hazard applications and is used for internal protection, for both indoor and outdoor applications. A wide variety of AVBs are produced for specific installations, such as lab faucets and equipment that use built-in AVBs.

Installation criteria. An AVB must be installed in the orientation designated by the approval agency recognized by the authority having jurisdiction. No control valves shall be installed downstream of the AVB device.

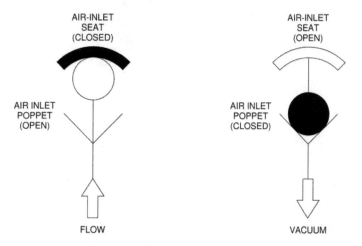

Figure 4-4   Atmospheric vacuum breaker
Courtesy of Apollo Valves/Conbraco Industries, Inc.

An AVB must not be subjected to conditions that would exceed its maximum working water pressure and temperature rating. A pipeline should be thoroughly flushed before an AVB is installed to ensure that no dirt or debris is delivered into the device. An AVB shall not be installed where it could be subjected to backpressure.

An AVB must not be installed in a pit or below grade where the air inlet could become submerged in water or where fumes could be present at the air inlet, as water or fumes may enter the device.

An AVB shall be installed a minimum of 6 in. (152 mm) above the highest point of use and any downstream piping coming from the device. Installation heights may vary. The local administrative authority should be consulted for optional acceptance criteria. An AVB shall not be subjected to continual use and shall not be pressurized for more than 12 hr in a 24-hr period.

Hose Connection Vacuum Breaker

Description. The hose connection vacuum breaker (HCVB) shall contain an internally loaded check valve biased to the closed position and an air-inlet valve biased to an open position when the device is not exposed to pressure. The HCVB is intended for installation on the discharge side of a faucet, hose bib, or hydrant fitted with hose threads. Plumbing codes may require a tamper-resistant means for attachment (i.e., breakaway screw). The HCVB shall be approved by an approval agency acceptable to the local administrative authority.

Function. As water enters the device inlet, the pressure will first cause the air-inlet valve to close (Figure 4-5). After the air inlet valve is sealed, the check valve will open, allowing water to flow into the downstream piping. During normal operation, the check valve will open in response to demand for water on the outlet and the air inlet will remain closed. When the demand for water ceases, the check valve will close.

In a backsiphonage condition, the inlet pressure will be reduced to subatmospheric, and the check valve will close as a result of higher downstream pressure. If the check valve fails to seal, the pressure will be relieved by the air-inlet opening, thereby breaking the vacuum and allowing air to be siphoned into the plumbing system instead of the downstream water.

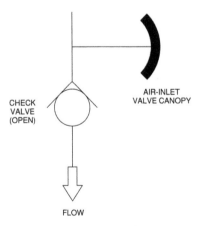

Figure 4-5 Hose connection vacuum breaker

Courtesy of Apollo Valves/Conbraco Industries, Inc.

Application. The HCVB is designed to prevent backflow from backsiphonage and low-head backpressure due only to elevated downstream piping, which should not exceed 10 ft (3.0 m) in height. The HCVB can be used for both low and high hazard applications and shall not be subjected to more than 12 continuous hours of pressure.

Installation criteria. The HCVB shall not be subjected to conditions that would exceed its maximum working water pressure and temperature rating.

Pressure Vacuum Breaker Assembly

Description. The pressure vacuum breaker (PVB) assembly shall contain an independently operating, internally loaded check valve and an independently operating, loaded air-inlet valve located on the discharge side of the check valve (Figure 4-6). In addition, the PVB assembly shall have an inlet and outlet resilient-seated, fully ported shutoff valve and two properly located resilient-seated test cocks. The PVB shall be installed as an assembly designed and constructed by the manufacturer with no field modifications being allowed. The PVB shall be approved by an approval agency acceptable to the local administrative authority.

Function. The check valve in the PVB is designed to generate a loading capable of holding a minimum of 1 psi (6.9 kPa) in the direction of flow with the outlet side of the check valve at atmospheric pressure. After water passes the check valve, it will cause the air-inlet poppet/valve to close by overcoming the air-inlet loading, which is designed to be a minimum of 1 psi (6.9 kPa). During normal operation, the check valve will open in response to demand for water on the downside and the air inlet will remain closed. When the demand for water ceases, the check valve will close.

In a backsiphonage condition, the inlet pressure will be reduced to a subatmospheric pressure (Figure 4-7). The check valve will close because of the higher pressure on the downstream side of the check valve. When the pressure on the downstream side of the check valve falls to the air-inlet opening point (minimum of 1 psi [6.9 kPa]), the air inlet will open to ensure that any vacuum is broken. If the check valve does not seal properly, the area after the check valve will decrease in pressure, causing the air-inlet poppet to come off the air-inlet seat; this action will open and break any vacuum by allowing air to be siphoned into the plumbing system instead of the downstream water.

Application. The PVB is an assembly designed to prevent backflow only from backsiphonage (Figure 4-7) and can be used for both high and low hazard applications. The PVB can be used for internal protection. Normally, the PVB is not used for service protection because of its inability to protect against backpressure. A PVB may be allowed for service protection in certain jurisdictions; however, restrictions may apply such as a dedicated, single-use service and supplying an irrigation system.

Installation criteria. A PVB must be installed in the orientation as approved by the agency recognized by the jurisdictional authority.

A PVB must not be subjected to conditions that would exceed its maximum working water pressure and temperature rating. The increased pressure that can happen from the creation of a closed system also must be evaluated. A PVB shall not be installed where it is subjected to backpressure.

A PVB should be sized hydraulically, taking into account both volume requirements and pressure loss through the assembly. A pipeline should be thoroughly flushed before a PVB is installed to ensure that no dirt or debris is delivered into the assembly.

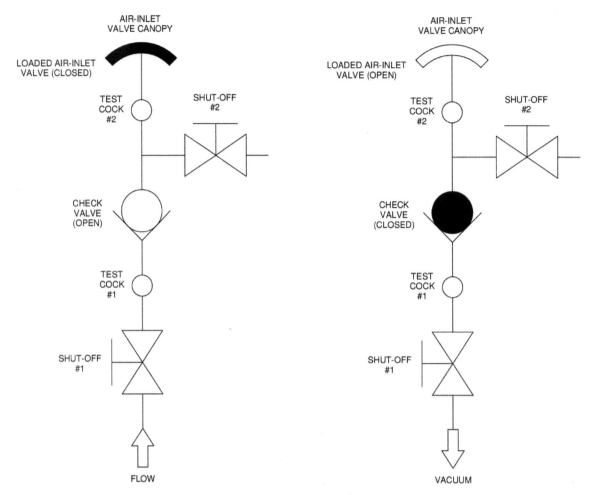

Figure 4-6 Pressure vacuum breaker assembly, normal flow condition

Courtesy of Apollo Valves/Conbraco Industries, Inc.

Figure 4-7 Pressure vacuum breaker assembly, backsiphonage condition

Courtesy of Apollo Valves/Conbraco Industries, Inc.

A PVB must not be installed in a pit or below grade where the air inlet could become submerged in water or where fumes could be present at the air inlet, allowing water or fumes to enter the assembly.

A PVB shall be installed a minimum of 12 in. (305 mm) above the highest point of use and any downstream piping supplied from the assembly. The installation should not be installed where platforms, ladders, or lifts are required for access. If an assembly must be installed higher than 5 ft (1.5 m) above grade, a permanent platform should be installed around the assembly. (Local administrative authorities should be consulted.)

A PVB shall be installed where it can be easily field tested and repaired as necessary. The assembly shall have adequate clearance around it to facilitate disassembly, repairs, testing, and other maintenance.

A PVB will periodically discharge water from the air inlet. The effect of this discharge on the area around the assembly must be evaluated. If a PVB is subjected to environmental conditions that could freeze or heat the assembly beyond its working temperatures, some means of protection should be installed to provide the correct temperature environment in and around the assembly.

Caution: When a pressure vacuum breaker is pressurized or repressurized, the air inlet will discharge water. Depending on the design, the discharge of gallons of water per minute may continue until the supply pressure reaches 2–7 psi. For this reason, the PVB assemblies are best suited for outdoor applications unless special attention has been paid to prevent water damage from the air-inlet water discharge.

Spill-Resistant Vacuum Breaker

Description. The spill-resistant vacuum breaker (SVB) shall contain an internally loaded check valve and a loaded air-inlet valve located on the discharge side of the check valve. In addition, the SVB assembly shall have an inlet and outlet, resilient-seated, fully ported shutoff valve and a properly located, resilient-seated test cock and vent valve. The SVB shall be installed as an assembly designed and constructed by the manufacturer with no field modifications allowed. The SVB shall be approved by an approval agency acceptable to the local administrative authority.

Function. The check valve is designed to generate a loading capable of holding a minimum of 1 psi (6.9 kPa) in the direction of flow of the check valve with the outlet side of the check at atmospheric pressure (Figure 4-8). As water enters the assembly inlet, the pressure will first cause the air-inlet poppet/valve to rise and seal against the air-inlet seat. After the air inlet is sealed, the check assembly will open, allowing water into the piping system. During normal operation, the check valve will open in response to demand for water on the outlet and the air inlet will remain closed. When the demand for water ceases, the check valve will close.

In a backsiphonage condition, the inlet pressure will be reduced to a subatmospheric pressure. The check valve will close and seal because of the higher pressure on the downstream side of the check valve. If the pressure in the SVB body is reduced by usage on the downstream side, the pressure will be relieved until the air-inlet poppet comes off the air-inlet seat and opens. If the check valve does not seal properly, the area after the check valve will decrease in pressure, causing the air inlet to open. This action will break any vacuum by allowing air to be siphoned into the plumbing system instead of the downstream water.

Application. The SVB is an assembly designed to prevent backflow only from backsiphonage. The SVB can be used for both high and low hazard applications. Normally, the SVB is authorized only for internal protection including both indoor and outdoor applications.

Installation criteria. An SVB must be installed in the orientation as approved by the agency recognized by the authority having jurisdiction.

An SVB must not be subjected to conditions that would exceed its maximum working water pressure and temperature rating. The increased pressure that can happen from the creation of a closed system also must be evaluated. An SVB shall not be installed where it is subjected to backpressure.

An SVB should be sized hydraulically, taking into account both volume requirements and pressure loss through the assembly. A pipeline should be thoroughly flushed before an SVB is installed to ensure that no dirt or debris is delivered into the assembly. An SVB must not be installed in a pit or below grade where the air inlet could become submerged in water or where fumes could be present at the air inlet, allowing water or fumes to be siphoned into the assembly.

An SVB shall be installed a minimum of 12 in. (305 mm) above the highest point of use and any downstream piping coming from the assembly. Installation heights may be varied. Local administrative authorities should be consulted for optional acceptance criteria. The SVB should not be installed where platforms, ladders, or lifts are required for access. If an assembly must be installed higher than 5 ft (1.5 m) above grade, a permanent platform should be installed around the assembly. (Local administrative authorities should be consulted for variances.)

50 BACKFLOW PREVENTION AND CROSS-CONNECTION CONTROL

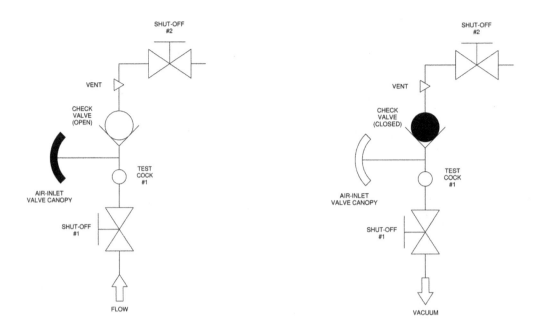

Figure 4-8 Spill-resistant vacuum breaker, normal flow and backsiphonage conditions
Courtesy of Apollo Valves/Conbraco Industries, Inc.

An SVB shall be installed where it can be easily tested and repaired as necessary. The assembly shall have adequate clearance around it to facilitate disassembly, repairs, testing, and other maintenance.

Double Check Valve Backflow Prevention Assembly

Description. The double check (DC) valve backflow prevention assembly shall contain two internally loaded, independently operating, approved check valves; two resilient-seated shutoff valves; and four properly located test cocks (Figure 4-9). The DC shall be installed as an assembly designed and constructed by the manufacturer. The assembly shall be approved by an approval agency acceptable to the local administrative authority.

Function. The check valves are designed to generate a loading capable of holding a minimum of 1 psi (6.9 kPa) in the direction of flow of the check valve, with the outlet side of the check valve at atmospheric pressure. During normal operation, the check valves will open in response to demand for water at the outlet (Figure 4-9). When the demand for water ceases, the check valves will close.

In a backpressure condition, the increase of pressure on the outlet will cause the second check to close tighter. If the second check does not seal properly, the first check will act as a backup to the second check (Figure 4-10). Foreign debris or deterioration of the check components can affect both checks simultaneously, rendering the double check incapable of preventing backflow without an outward indication of failure.

In a backsiphonage condition, the inlet pressure will be reduced to a subatmospheric pressure. The greater pressure on the downstream side of the second check will cause the second check to close. If the second check does not seal off properly, the first check will act as a backup (Figure 4-11).

Application. The DC is an assembly designed to prevent backflow from backpressure and/or backsiphonage. A DC can be used only for low hazard applications and for service protection or internal protection (Figure 4-12).

AWWA Manual M14

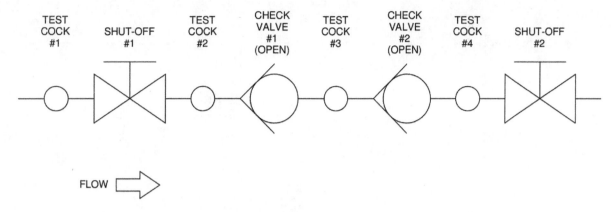

Figure 4-9 Check valves open, permitting flow
Courtesy of Apollo Valves/Conbraco Industries, Inc.

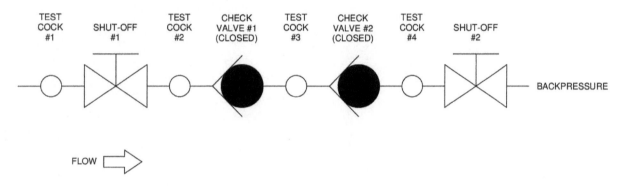

Figure 4-10 Backpressure, both check valves closed
Courtesy of Apollo Valves/Conbraco Industries, Inc.

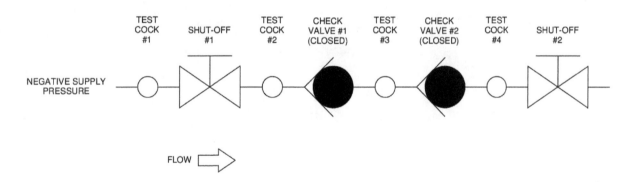

Figure 4-11 Negative supply pressure, check valves closed
Courtesy of Apollo Valves/Conbraco Industries, Inc.

Installation criteria. A DC must be installed in the orientation as it was approved by the approval agency recognized by the jurisdictional authority with no field modifications allowed. A DC must not be subjected to conditions that would exceed its maximum working water pressure and temperature rating. The increased pressure that can happen from the creation of a closed system also must be evaluated to prevent damage to the assembly or other plumbing-system components.

52 BACKFLOW PREVENTION AND CROSS-CONNECTION CONTROL

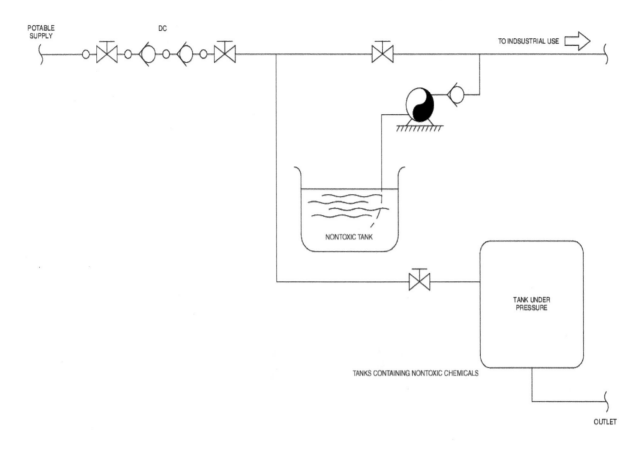

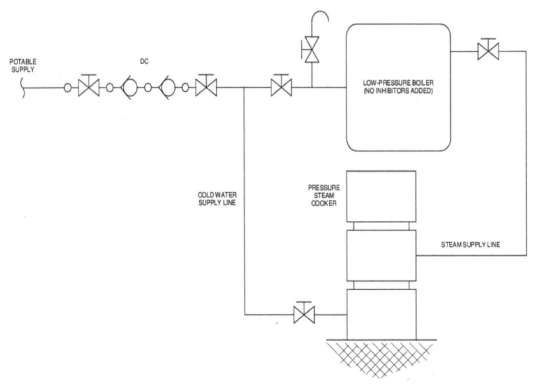

Figure 4-12 Typical double check valve assembly applications
Courtesy of Apollo Valves/Conbraco Industries, Inc.

A DC should be sized hydraulically, taking into account both volume requirements and pressure loss through the assembly. A DC should not be installed in a pit or below grade because of potential for flooding, confined space entry, and other safety concerns.

A pipeline should be thoroughly flushed before a DC is installed to ensure that no dirt or debris is delivered to the assembly. The DC should not be installed where platforms, ladders, or lifts are required for access. If an assembly must be installed higher than 5 ft (1.5 m) above grade, a permanent platform shall be installed around the assembly. (Local administrative authorities should be consulted.)

A DC shall be installed where it can be easily field tested and repaired as necessary. The assembly shall have adequate clearance around it to facilitate disassembly, repairs, testing, and other maintenance. If a DC must be subjected to environmental conditions that could freeze or heat the assembly beyond working temperatures, protection should be installed to provide the correct temperature environment in and around the assembly.

Reduced-Pressure Principle Backflow Prevention Assembly

Description. A reduced-pressure principle (RP) backflow prevention assembly shall contain two internally loaded, independently acting check valves with a hydraulically operating, mechanically independent differential pressure relief valve located between the check valves and below the first check valve. The check valves and the relief valve shall be located between two tightly closing, fully ported, resilient-seated shutoff valves. The RP shall have four properly located resilient-seated test cocks, as shown in Figure 4-13. The RP shall be installed as an assembly designed and constructed by the manufacturer with no field modifications allowed. An RP shall be approved by an approval agency acceptable to the local administrative authority.

Function. An RP is designed to maintain a pressure that is lower after the first check than it is at the inlet. The water pressure into the RP will be reduced by the amount of the first check loading (a minimum of 3 psi [20.7 kPa] higher than the relief-valve opening point). When the pressure is reduced after the first check, the relief valve senses the difference between the inlet pressure (before the first check) and the pressure after the first check. The relief valve ensures that the pressure after the first check is always lower than the inlet pressure by the amount of the relief-valve opening point, which shall be a minimum of 2 psig (13.8 kPa gauge). If the pressure in the area after the first check increases to within a minimum of 2 psig (13.8 kPa gauge) less than the inlet pressure, the relief valve will open to ensure that a lower pressure is maintained. The second check is located downstream from this relief valve and will reduce the pressure by the amount of the check loading, which is a minimum of 1 psi (6.9 kPa). In a normal flowing situation, both check valves will be open to meet the demand for water and the relief valve will stay closed. When the demand for water ceases, both checks will close and the relief valve will stay closed. See Figures 4-13 and 4-14.

In a backpressure condition, both check valves will close, and the second check will stop the increased pressure from traveling into the area between the two checks (Figure 4-15). If the second check is not maintaining its separation of pressure, the backpressure will leak past the second check and cause the pressure in the area between the two checks to increase. After the increase in pressure rises to the inlet pressure less the relief-valve opening point (minimum of 2 psig [13.8 kPa gauge]), the relief valve will open and discharge water from the assembly to the atmosphere. This discharge from the relief valve ensures that the pressure after the first check is always lower than the inlet pressure.

In a backsiphonage condition, the inlet pressure will be reduced to a subatmospheric pressure, and the pressure downstream from the first check will cause the relief valve to open and discharge the water to the atmosphere (Figure 4-16).

Application. The RP is an assembly that can prevent backflow from backpressure and/or backsiphonage. An RP is designed for both high and low hazard applications and

54 BACKFLOW PREVENTION AND CROSS-CONNECTION CONTROL

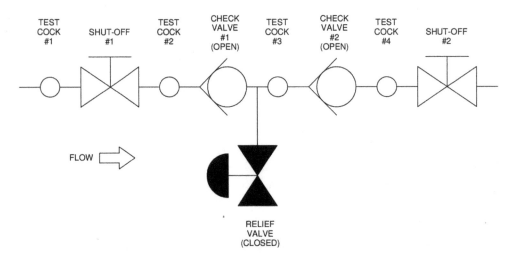

Figure 4-13   Reduced-pressure principle backflow prevention assembly, both check valves open and the differential relief valve closed

Courtesy of Apollo Valves/Conbraco Industries, Inc.

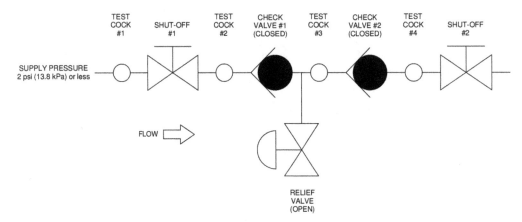

Figure 4-14   Both check valves closed and the differential pressure relief valve open can be used for service protection or internal protection (see Figure 4-17)

Courtesy of Apollo Valves/Conbraco Industries, Inc.

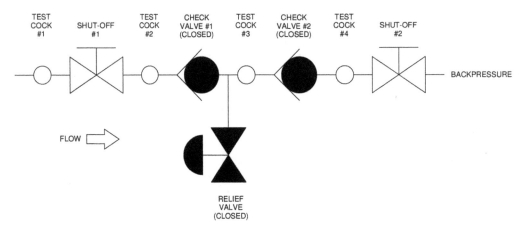

Figure 4-15   Backpressure: both check valves closed and the differential pressure relief valve closed

Courtesy of Apollo Valves/Conbraco Industries, Inc.

AWWA Manual M14

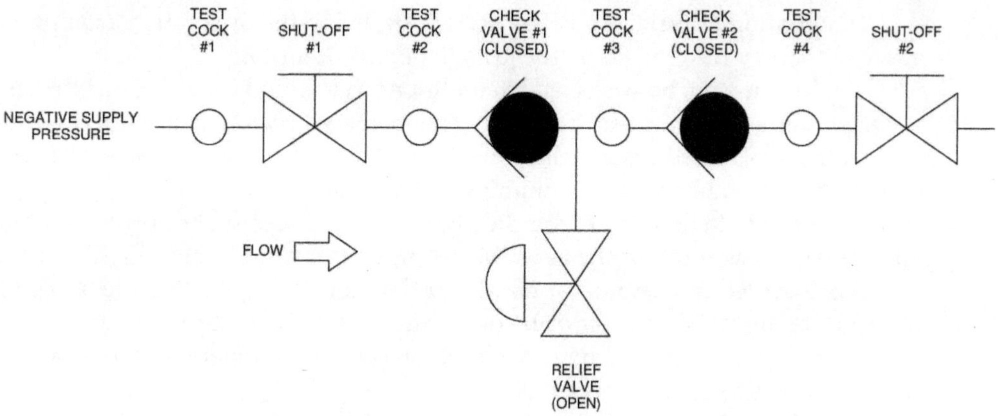

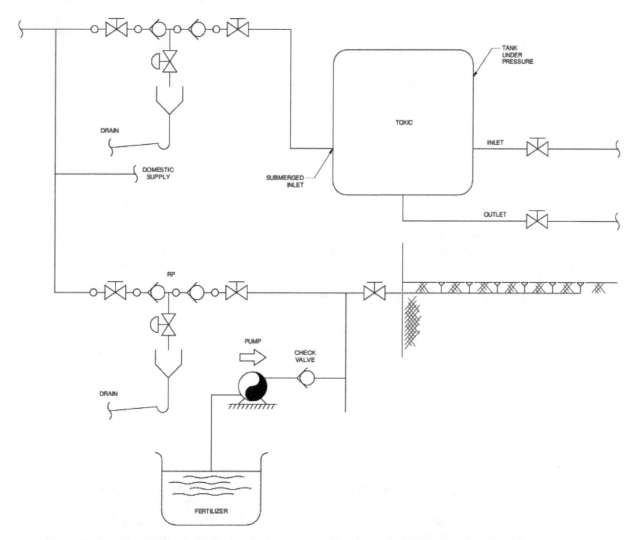

Figure 4-16   Backsiphonage: both check valves closed and the differential pressure relief valve open

Courtesy of Apollo Valves/Conbraco Industries, Inc.

Figure 4-17   Typical reduced-pressure principle backflow prevention application

Courtesy of Apollo Valves/Conbraco Industries, Inc.

Installation criteria. An RP must be installed in the orientation as approved by the approval agency recognized by the jurisdictional authority.

An RP must not be subjected to conditions that would exceed its maximum working water pressure and temperature rating. The increased pressure that can occur because of the creation of a closed system also must be evaluated because excessive backpressure can damage the assembly or other plumbing components.

An RP should be sized hydraulically, taking into account both volume requirements and pressure loss through the assembly. A pipeline should be thoroughly flushed before an RP is installed to ensure that no dirt or debris is delivered into the assembly. An RP must not be installed in a pit or below grade where the relief valve could become submerged in water or where fumes could be present at the relief-valve discharge, allowing water or fumes to enter the assembly.

An RP shall be installed a minimum of 12 in. (305 mm) above the relief-valve discharge-port opening and the surrounding grade and floodplain (as regulated). The installation should not be installed where platforms, ladders, or lifts are required for access. If an assembly is installed higher than 5 ft (1.5 m) above grade, a permanent platform should be installed around the assembly. (Local administrative authorities should be consulted to confirm acceptability.)

An RP shall be installed where it can be easily tested and repaired. The assembly shall have adequate clearance around it to facilitate disassembly, repairs, testing, and other maintenance.

An RP may periodically discharge water from the relief valve. The effect of this discharge must be evaluated. If the RP discharge is piped to a drain, an air-gap separation must be installed between the relief-valve discharge opening and the drain line leading to the drain. Most RP manufacturers produce an air-gap drain fitting that attaches to the relief valve so that piping can run from the RP to a drain. *Caution:* The air-gap drain is designed to carry occasional small amounts of discharge and/or spitting from the relief valve. The full discharge of an RP generally is beyond the capacity of the air-gap drain fitting. The appropriate drainage should be sized to prevent water damage per the manufactures installation instructions.

If an RP must be subjected to environmental conditions that could freeze or heat the assembly beyond its working temperatures, some means of protection should be installed to provide the correct temperature environment in and around the assembly.

Double Check Detector Backflow Prevention Assembly

Description. The double check detector backflow prevention assembly (DCDA) shall consist of a main-line DC with a bypass (detector) arrangement around the main-line DC that shall contain a water meter and a DC (Figure 4-18). The DCDA shall be installed as an assembly designed and constructed by the manufacturer. The bypass arrangement may bypass the second main-line check only, and contain a water meter and a single check (Figure 4-19). This is commonly referred to as a Type II DCDA and it provides the same level of backflow protection as a DCDA but uses a shared first check valve for both the mainline and bypass arrangements.

Function. The DCDA operates like a DC except the bypass is engineered to detect the first 2 gpm (7.6 L/min) of flow through the assembly. This low flow is registered by the water meter in the bypass and is used to show any unauthorized usage or leaks in the fire protection system.

Application. The DCDA and Type II DCDA, as seen in Figures 4-18 and 4-19, are designed for fire-protection systems for which a main-line meter is not used but leaks or unwanted usage are detected in the bypass. A DCDA or Type II DCDA can protect against backpressure and/or backsiphonage and be used only for lowhazard service protection or internal protection applications.

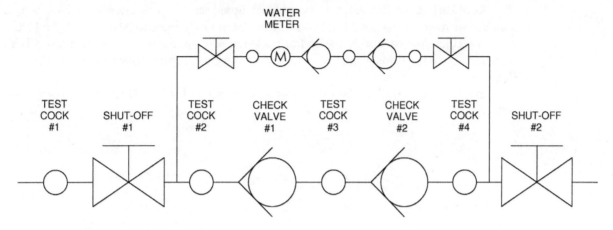

Figure 4-18 **Double check detector backflow prevention assembly**
Courtesy of Apollo Valves/Conbraco Industries, Inc.

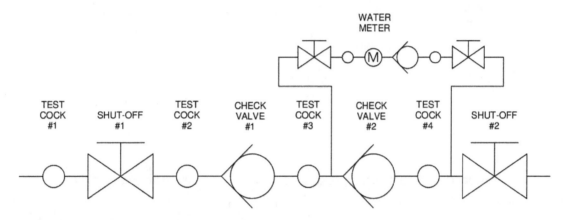

Figure 4-19 **Type II double check detector backflow prevention assembly**
Courtesy of Apollo Valves/Conbraco Industries, Inc.

Installation criteria. The criteria are the same as for the DC (see DC installation criteria).

Reduced-Pressure Principle Detector Backflow Prevention Assembly

Description. The reduced-pressure principle detector backflow prevention assembly (RPDA) shall consist of a main-line RP with a bypass arrangement around the RP that shall contain a water meter and an RP. The RPDA shall be installed as an assembly designed and constructed by the manufacturer. The bypass arrangement may bypass the second main-line check only, and contain a water meter and a single check. This is commonly referred to as a Type II RPDA, and it provides the same level of backflow protection as an RPDA, but uses a shared first check valve for both the main-line and bypass arrangements.

Function. The RPDA and Type II RPDA, as seen in Figures 4-20 and 4-21, operate like an RP except the bypass is engineered to detect the first 2 gpm (7.6 L/min) of flow through the assembly. This low flow is registered by the water meter in the bypass and is used to show any unauthorized usage or leakage in the fire protection system.

Application. An RPDA is an assembly designed for fire-protection systems in which a main-line meter is not used but leaks or unwanted usage need to be detected. An RPDA can protect against backpressure and/or backsiphonage and can be used to protect both high and low hazard installations applications. An RPDA can be used for service protection or internal protection.

Installation criteria. The criteria are the same as for the RP (see RP installation criteria).

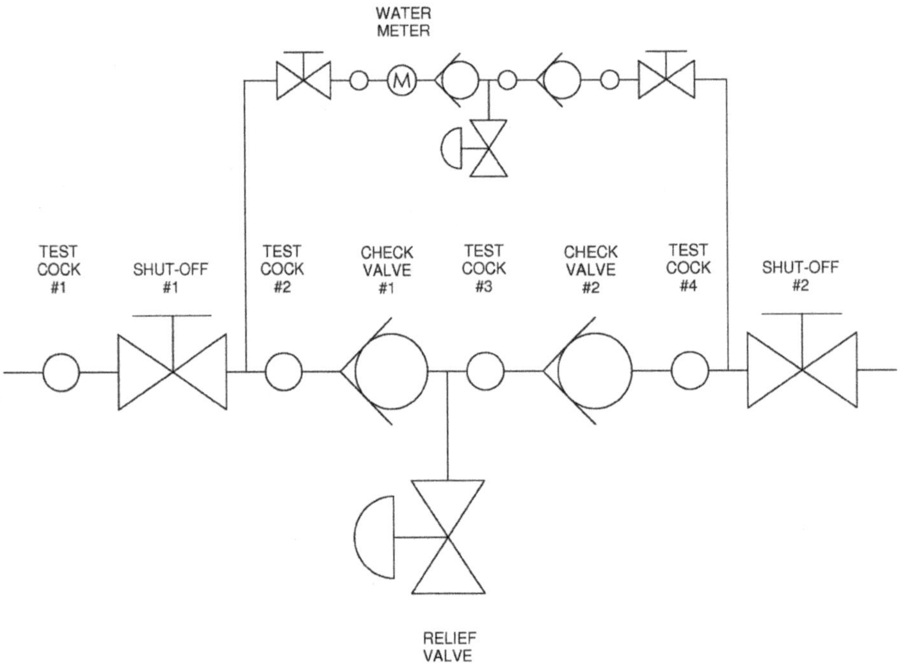

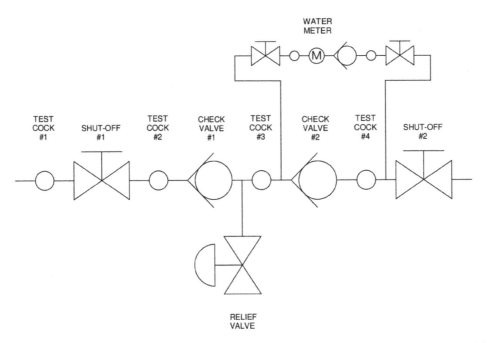

Figure 4-20   Reduced-pressure principle detector backflow prevention assembly

Courtesy of Apollo Valves/Conbraco Industries, Inc.

Figure 4-21   Type II reduced-pressure principle detector backflow prevention assembly

Courtesy of Apollo Valves/Conbraco Industries, Inc.

METHOD FOR CONTROLLING BACKFLOW

Air Gap

Description. A proper air gap (AG) is an acceptable method to prevent backflow; however, the disadvantage is that supply-system pressure is lost. An approved AG is a piping system that provides an unobstructed vertical distance through free atmosphere between the lowest point of a water supply outlet and the overflow rim of an open, nonpressurized receiving vessel into which the outlet discharges. For AGs that are constructed on site, these vertical physical separations must be at least twice the effective opening (inside diameter) of the water supply outlet but never less than 1 in. (25 mm). In locations where the outlet discharges within three times the inside diameter of the pipe when measured *from a single wall* or other obstruction, the AG must be increased to three times the effective opening but never less than 1.5 in. (38 mm). In locations where the outlet discharges within four times the inside diameter of the pipe when measured *from two intersecting walls*, the AG must be increased to four times the effective opening but never less than 2 in. (51 mm).

Air gaps should not be approved for locations where there is the potential for the atmosphere around the AG to be contaminated, nor should the inlet pipe be in contact with a contaminated surface or material. Air gaps are often designed as an integral part of plumbing fixture and equipment and shall meet the requirements of ASME A112.1.2 Standard.

Application. An AG can be used for service or internal protection (Figures 4-22 through 4-23). A properly installed and maintained AG is the best means to protect against backflow because it provides a physical separation between the water source and its use. In a backsiphonage condition, an AG can allow the surrounding atmosphere to enter the piping system. The AG must be located in such a way to ensure that fumes or other airborne substances cannot be siphoned into the potable water system. The installation shall not include any interference with the free-flowing discharge into the receiving vessel. No solid material shields or splash protectors can be installed. Screen or other perforated material may be used if it presents no interference. AGs must be inspected at least annually to ensure that the proper installation is maintained and has not been circumvented.

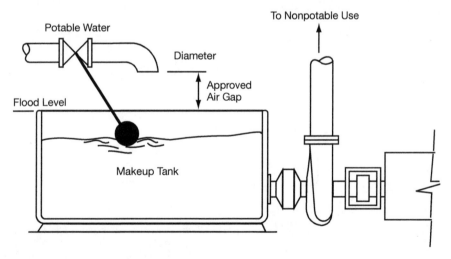

Figure 4-22   AG on tank

60 BACKFLOW PREVENTION AND CROSS-CONNECTION CONTROL

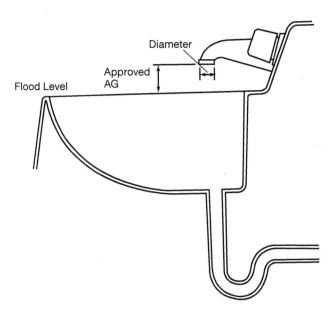

Figure 4-23   AG on lavatory

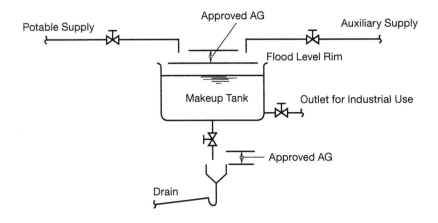

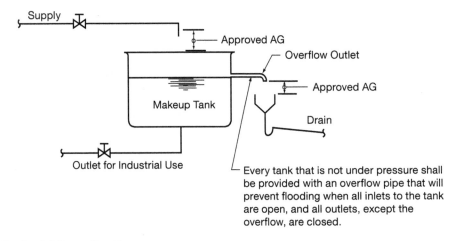

Figure 4-24   Typical AG applications

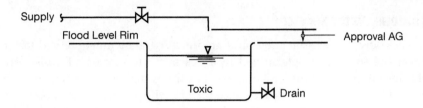

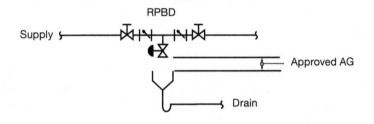

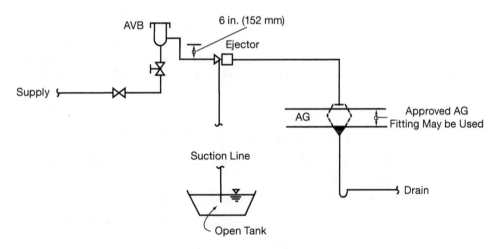

Figure 4-25   Additional typical AG applications

FIELD TESTING

Backflow Prevention Assembly Tester

A field tester is a person who has demonstrated competence to test, repair, and troubleshoot backflow prevention assemblies as evidenced by certification that is recognized by the authority having jurisdiction.

Continuous Water Service

When backflow prevention assemblies are installed, they must be field tested and maintained at the frequency determined by the water supplier or administrative authority, but not less than annually. This means that the water must be shut off temporarily. If the assembly is feeding a critical service facility where water cannot be interrupted for a period of time, more than one assembly should be installed in a parallel installation. In this way, one assembly can be field-tested or repaired while the other assembly delivers water to the facility. Two or more parallel assemblies shall be the same type of assembly, as the degree of hazard shall dictate. When parallel assemblies are used, the piping system should be sized hydraulically, taking into account both volume requirements and pressure loss through the assemblies. Each assembly shall operate normally.

Testing Awareness

Backflow prevention assemblies are installed in locations where a reliable means of backflow protection is deemed necessary. To ensure the continued protection of an identified actual or potential cross-connection, the properly installed backflow prevention assemblies must be field tested at least annually to ensure that they continue to prevent backflow. If, in the testing process, the assembly is found to be working at less than its minimum criteria, it shall be repaired and returned to its full ability to prevent backflow.

A person shall be allowed to field test backflow prevention assemblies only after successfully completing a training course and passing a certification course for backflow prevention assembly testers. Some jurisdictions may require additional business and/or performance licensing to fulfill the duties of a backflow prevention assembly tester.

Backflow prevention assemblies may need to be field tested more than annually. In addition to annual field testing, backflow prevention assemblies are field tested

- immediately following initial installation
- whenever an assembly is relocated
- whenever supply piping is altered
- whenever an installed assembly is newly discovered and previous testing records are not available
- whenever an assembly is taken apart for repair
- whenever the administrative authority requires more frequent testing to ensure continued protection
- whenever the assembly is taken out of service and is returned to service (does not include loss of distribution system pressure)

Field-Testing Procedures

The field-test procedure collects accurate data about the workings of a backflow prevention assembly. The data collected from the field test must be compared to minimum acceptable standards to ensure that the backflow prevention assembly is performing properly and preventing backflow. If the data from the field test shows that the backflow prevention assembly does not meet the minimum criteria, the assembly shall be repaired or replaced.

Because of the many variations in specific, detailed field-test procedures and because various regulatory authorities accept different field-test procedures, this manual will not identify any specific procedure as acceptable in lieu of other locally approved procedures. (See appendix B for various sample field-test procedures, which are provided for information only and are not an endorsement by AWWA.) The local administrative authority shall evaluate available field-test procedures and specify acceptable procedure(s) for each type of assembly that will be used by testers in their jurisdiction.

The administrative authority must be assured that testers in its jurisdiction can properly perform field-test procedures on backflow prevention assemblies. To provide this assurance, each tester must successfully complete a recognized examination process. This examination shall include testing about the theory of backflow prevention and a hands-on, practical examination of field test procedures on all types of backflow prevention assemblies in both properly working and malfunctioning conditions. The successful completion of this examination results in certification that allows the tester to perform field tests of backflow prevention assemblies and to submit field-test data to the administrative authority. The certification shall be valid for a specific length of time before recertification is required.

After the certified tester has properly performed the field-test procedure on a backflow prevention assembly, the data collected shall be recorded on a field-test report form. Jurisdictions may allow the field test data to be recorded electronically. The data on this field-test report form shall properly represent the workings of the assembly at the time it was tested by the certified tester. Copies of the field-test report form should be sent to all appropriate interested parties including, as a minimum, the water user and the local administrative authority. The tester shall retain copies of the test data. Records must be maintained for a time period specified by the authority having jurisdiction.

Field-Test Equipment

To collect accurate data in the field about the operation of a backflow prevention assembly, a certified tester must adhere to a proper field-test procedure. The field-test equipment is used to test backflow prevention assemblies and must be capable of providing accurate data. There are many types of commercially available field-test equipment. The water supplier or administrative authority must verify that the field-test equipment used can perform the field-test procedure that is accepted in its jurisdiction and that the accuracy of the equipment is maintained. To ensure the ability of the test equipment to provide accurate data, the accuracy of the field-test equipment shall be verified at least annually. If the accuracy of the field-test equipment is not within accepted standards recognized by the local administrative authority, the field-test equipment shall be calibrated and brought into acceptable accuracy tolerances.

Maintenance of assemblies. When a certified tester performs an accurate field test on an approved backflow prevention assembly with accurate field-test equipment, the data shall properly represent the working abilities of the backflow preventer. When this data indicates that the assembly is not performing to the minimum accepted standards as established by the field-test procedure, the results shall be noted on the field-test report, and the assembly shall be repaired or replaced to ensure that the assembly will continue to prevent backflow. The purpose of a repair procedure is to return the assembly to a condition in which it will reliably protect against backflow. All backflow prevention assemblies are designed to be repaired. Only authorized original factory repair parts as specified by the manufacturer should be used.

The first step in the repair process should be to obtain the necessary repair information. Most manufacturers provide information about the repair processes for the assemblies they produce. New, original factory repair parts should be obtained. The use of recycled components or flipping check disks is not an acceptable practice because the assembly was evaluated and approved with only new components. The repair process will involve the disassembly of the backflow preventer and the proper placement of the new repair parts. Care must be taken to safely perform the repair. After the repair is performed, the backflow prevention assembly must be field tested by a certified tester to ensure that the repair procedure has restored the assembly to proper working condition.

It may not be possible to repair some backflow prevention assemblies. This usually involves assemblies that do not have replaceable parts, such as check seats, that may be cast into the body of some of the older assemblies. The original manufacturer may no longer produce repair parts for some older assemblies. In these cases, the assembly may need to be replaced. When replacing an existing assembly with a new one, the application should be reviewed to ensure that the proper type of backflow prevention assembly is installed for that application.

Tester Responsibilities

Accurate data about the workings of properly installed (and approved) backflow prevention assemblies are important to the administration of an effective cross-connection control program. The tester is responsible for the accurate generation of data, a correct assessment of the workings of each assembly tested, and proper dissemination of the data to all necessary parties. Certified testers shall adhere to the field-test procedures established in the certification process. Local administrative authorities shall review the data for validity, and they may elect to establish local criteria in addition to the procedures established by the accepted certification program. If local administrative authorities choose to establish additional criteria, these criteria must be communicated to testers. A periodic meeting or other contact with testers can help to clarify the requirements.

REFERENCE

American Society of Mechanical Engineers (ASME). 2012. *Air Gaps in Plumbing Systems (for Plumbing Fixtures and Water-Connected Receptors)*. New York: ASME.

AWWA MANUAL

M14

Chapter 5

Conducting a Cross-Connection Control Survey

Cross-connection control surveys (herein referred to as surveys) are necessary to protect the public water system and water consumers. To ensure that the water distribution system and plumbing remain safe for all users, a survey is required by many states and local jurisdictions. A survey is an evaluation of a residential, commercial, industrial, or institutional facilities' plumbing system to assess the degree of hazard of each cross-connection. Where residential connections are not included in the survey program, a public education program targeted toward residential cross-connections is essential. The surveys should be conducted by a person trained in cross-connection control. The person conducting the survey should have a thorough understanding of the adopted plumbing code and the local cross-connection control program to ensure the inspection performed incorporates the local program's approach to providing backflow protection (i.e., premises containment protection, internal/fixture protection, or a hybrid of combining both program approaches). Program administration is discussed in chapter 3.

A *cross-connection* is defined as the actual or potential physical connection of a safe or potable water supply with another water source of unknown or contaminated quality possibly causing the potable water to be contaminated or polluted. *Backflow* is defined as a hydraulic condition, caused by a difference in pressures, that causes nonpotable water or another substance to flow into a potable water system. Cross-connection surveys are included in the United States Environmental Protection Agency's (USEPA) Revised Total Coliform Rule (USEPA 1999) response toolkit because a possible source of microbial contamination is backflow.

A domestic or potable water supply system must be designed, installed, and maintained so that nonpotable water is unable to enter the potable water supply. All plumbing fixtures used for potable purposes are required to be free of cross-connections or equipped with an approved backflow prevention device or assembly.

AUTHORITY AND RESPONSIBILITIES

The Safe Drinking Water Act (1974, as amended) gave the USEPA the authority to establish national standards for drinking water. In Canada, acts and regulations vary from province/territory to province/territory. The public water system is required to be in compliance with these standards through state drinking water programs. These include warranties that the drinking water that is delivered and used for potable purposes conforms to the standards established by the USEPA or provincial/territorial ministry. When considering compliance with these standards, the public water system should utilize a multiple barrier approach. This approach includes source protection, treatment optimization, and distribution protection that should include a disinfectant residuals and a cross-connection control program. The public water system should consider controlling cross-connections as an integral part of the public water system overall compliance program.

The role of the water supplier or individual conducting the survey is to ensure that the water quality in the water distribution system is maintained in compliance with all applicable acts and regulations, including the prevention of system contamination or pollution from backflow. The water supplier or individual conducting a survey is responsible for notifying the property owner of the results of the survey. The property owner is responsible for correcting any unprotected cross-connections by using qualified personnel. It should be noted that it is not the responsibility of the water supplier to recommend corrective actions to control a cross-connection but to only disclose the presence of potential or actual cross-connections found as a result of the survey and to require service/meter protection if needed. The water supplier is responsible for informing or educating all customers of the dangers associated with unprotected cross-connections. This may include using web content, special mailings, bill inserts, or information in the USEPA required annual water quality report to consumers (CCR).

PURPOSE OF A CROSS-CONNECTION CONTROL SURVEY

The purpose of a survey program is the protection of public health and includes one or more of the following:

- Identification of the degree of hazard
- Identification of where a closed piping system will be created by the installation of backflow preventers, and where provisions for thermal expansion should be considered
- Identification of existing backflow protection
- Identification of hydraulic conditions of actual and potential cross-connections
- When required by local ordinance, identification of the type of backflow protection required at the point of service
- When required by local ordinance, assistance in identifying the type of protection required to eliminate or control unprotected cross-connections where they exist

ASSESSING THE DEGREE OF HAZARD

Surveys are performed to identify actual and potential, controlled and uncontrolled cross-connections to the potable water system. Cross-connections present a general degree of hazard, which is based on the risk posed to the potable water supply. The degree of hazard is characterized as either a contaminant or pollutant. A contaminant is considered a high health hazard, and a pollutant would be considered a low, nonhealth hazard. Canadian provincial codes use the classifications of minor, moderate, and severe hazard.

Additional considerations used in assessing the degree of hazard may include but are not limited to

- Complexity of piping systems
- Internal protection
- Access to secondary and auxiliary water systems
- Likelihood of piping changes
- Ability of the water supplier to frequently monitor piping changes
- When required by the cross-connection control program or local ordinance

Where a premises isolation/containment protection program is implemented, the type of backflow protection required at the point of service is based on the degree of hazard established by one or more of the following:

- Type of facility as determined by a field inspection
- Hazards found within a facility
- Inability to have free and unrestricted access to the facility
- Multiple service connections that present the risk of a flow-through condition
- Potential for plumbing modifications and/or changes in use

Where an internal/fixture protection program is implemented, the person conducting the survey must assess the degree of hazard for each cross-connection found within the facility. A backflow protection recommendation or requirement must be made for each cross-connection.

SURVEY CONSIDERATIONS AND CONCEPTS

When conducting the cross-connection control survey, the individual conducting the survey should take into consideration and identify the following:

1. Actual or potential cross-connections between the potable and nonpotable water system
2. Hydraulic conditions associated with the cross-connection (backpressure or backsiphonage)
3. Impact on water quality from the associated degree of hazard
4. Proper backflow protection or action required to eliminate the individual cross-connection
5. Ensure containment at the point of water service connection

The individual conducting the survey should also make sure the owner has the right method to control the cross-connection. Evaluations should be made as to the appropriateness of the device or assembly protecting the potable water system. When field testing of assemblies occurs should also be evaluated for conformance with jurisdictional requirements.

The water supplier should consider several factors when requiring the installation of backflow preventers, including but not limited to the following:

- Hydraulics of the public water distribution system
- Pressure losses within the building
- Practical location of devices
- Thermal expansion
- Safe discharge of water

Service Protection Approach

Service protection (premise isolation or containment) is the installation of a backflow prevention device/assembly between the facility and public water distribution system. The backflow prevention device/assembly installed as service line protection ensures water of questionable quality cannot leave a facility and enter the public distribution system.

Where service line protection is utilized, the water supplier or cross-connection control surveyor should notify the owner of the facility of the potential for hazards within the building and make recommendations to correct unprotected cross-connections. The water supplier or the cross-connection control surveyor should also notify the appropriate plumbing code or local health official of the situation. The local ordinance or code may also require the facility owner or agent to conduct a survey of the internal plumbing system downstream of the service line protection.

Dedicated Water Line Approach

The dedicated water line approach consists of isolating the potable water line that originates after the water meter and any required service protection. This dedicated line is used solely for human consumption. Internal zone isolation or backflow protection installed at each fixture or appliance will prevent nonpotable water or process water from entering the dedicated water line. The adopted plumbing code may require color coding and regular (at least annual) inspections for unprotected cross-connections. A certificate of compliance may be issued by the authority having jurisdiction for required code compliance.

The type of containment device shall be commensurate with the degree of highest hazard found in the facility. All water after the containment backflow prevention device shall be considered nonpotable or process water. There should be no potable water lines installed on the nonpotable water line after the containment device. Once the nonpotable water line is established by the containment device, the water line must be color coded as required by the plumbing code. The installation of backflow prevention after the containment device is not required for public health protection but may be installed for hydraulic control (preventing cross-contamination between industrial processes). Zone isolation consists of the installation of a backflow preventer on a common feed to multiple points of nonpotable use.

When indoor installation of backflow preventers that discharge water (such as but not limited to AVB, PVB, RP) is necessary, provisions must be made to properly dispose of water in accordance with the plumbing code and any environmental discharge requirements.

Request for Internal Cross-Connection Control Information

Prior to conducting a survey, an initial interview of the facility owner or owner's designee must be performed. The surveyor must gather information sufficient to provide a complete understanding of the facility. Unrestricted access to the entire facility must be obtained.

CONDUCTING A CROSS-CONNECTION CONTROL SURVEY 69

A site employee that has full access and knowledge of the facility should be assigned to support the surveyor. All tenants and employees should be notified of the need for unrestricted access. Critical processes must be identified prior to any interruption to the water supply during shut-down testing. Other topics to discuss and items to obtain during this interview include but are not limited to

- Documentation of routine inspections of plumbing changes
- Previous surveys/inspection reports
- Schematics and mechanical drawings of the facility
- Number of water services and water meters, including serial numbers and locations
- Details of fire-suppression systems
- Location of auxiliary water (ponds, cisterns, storage tanks, etc.)
- Location of or plans to install water treatment processes (reverse osmosis (RO) systems, etc.)
- All uses of water (process, fill stations, domestic, etc.)
- Details of irrigation systems
- Heating and cooling systems
- Chemical systems (storage, handling, and delivery)
- Complete property owners' contact information
- Knowledge of local jurisdictional requirements
- Authorization letter from water supplier
- Permission to test all overdue existing assemblies (as applicable)

To facilitate a comprehensive survey, the person conducting the survey must also be aware of other factors including

- Size and complexity of the facility
- Approximate time required to complete the survey
- Presence of complex or multiple piping systems
- Inaccessible or concealed piping systems
- Inadequate piping identification
- Changes in plumbing configurations on a regular frequency
- Manufacturing or industrial fluids in piping systems for facility operations
- No current as-built/engineering drawings of the potable water system

CONDUCTING THE SURVEY

Surveys should be scheduled with the facility owner or owner's designee in advance. The survey should be conducted at a reasonable time to reduce any disruption to normal operations at the facility. If after proper notice the surveyor is refused access to the building or if customer plumbing is untraceable, the individual conducting the survey shall assume a cross-connection is present and take the necessary action to ensure the public water system is protected.

A survey consists of entering a facility from the point of water service entry (usually the meter) and identifying the piping and water use to each end point of use. During the course of the survey, it is the responsibility of the surveyor to

- Identify and note the location of actual or potential cross-connections
- Determine the degree of hazard (high or low)
- Determine if the cross-connection is direct or indirect
- Determine if continuous pressure is present (downstream control valves)
- Determine other hydraulic considerations such as but not limited to
  - pressure fluctuations
  - water hammer
  - pressure loss
  - thermal expansion
  - potential for backpressure and/or backsiphonage
  - discharge of water
- Evaluate test records for all backflow prevention assemblies
- Determine access requirements and if personal protection equipment is required to access the facility

Specialized Tools

The person conducting a survey will need tools and equipment to access, observe, and document details. Some equipment used by surveyors includes

- Proper identification (name badge or security clearance)
- Clipboard capable of holding and protecting forms and documents
- Copy of adopted plumbing code, regulations, and local ordinance
- Survey forms
- Copy of testing records and previous surveys and piping drawings
- List of approved assemblies and devices
- Waterproof pen, pencils, and eraser
- Tape measure or ruler
- Flashlights, standard and flexible inspection lights
- Laser pointer for tracking distant pipes
- Nonreversing inspection mirror
- Shatter-resistant mirror for looking around corners and over obstructions
- Small wire brush for cleaning nameplates
- Screwdrivers, standard and tamper-proof
- Binoculars
- Digital camera, still and/or video
- Digital voice recorder
- Personal protection equipment (hard hat, safety glasses, first-aid kit, hearing protection, air monitor, safety vest, gloves, coveralls)

When conducting the survey, the surveyor should consider dividing the survey into three groups—cross-connections located within the (1) public water systems, (2) internal domestic or potable water systems, and (3) fire-protection systems. Some water suppliers do not have jurisdiction or responsibility for cross-connection control on private premises. The surveyor should confirm that state, provincial, or local regulations provide the necessary authority.

Water distribution system. The public water system consists of the water distribution system up to the service connection or water meter to the location being served by the public water system. The water supplier is responsible for controlling cross-connections within the water distribution system. A survey of the distribution system, including the review of water treatment and distribution system (pump stations, blowoffs, air relief, bypasses, emergency connections, etc.), and a risk assessment of all water customers within the public water system, must be conducted. The highest priority for risk assessments should be placed on facilities that pose a high degree of hazard, that have a high probability that backflow will occur, or are known/suspected to have cross-connections. Cross-connections are controlled through the installation of backflow protection at the point of water service.

Domestic or potable water system. Where the identification of cross-connections within the domestic or potable water system located downstream of the service connection or water meter, surveys should be conducted to the terminal end. The survey of a potable water system should be conducted in coordination with the local code enforcement agency. Cross-connections are controlled by the installation of backflow protection as required by the plumbing code or local ordinance. Once an initial survey is completed, a re-survey timetable should be determined by the water supplier for each customer based on the degree of hazard and potential for backflow. The water supplier should periodically (timetable established by the authority having jurisdiction) re-inspect or require the owner to have the potable water system inspected for cross-connections to identify new, unprotected cross-connections. Other factors, such as new construction; change in zoning, ownership, or tenant; water quality complaints; or anomalies in water usage; may prompt an immediate re-survey.

Fire-protection system. The fire-protection system must be equipped with backflow protection commensurate with the degree of hazard within the system. A containment device should be installed on a fire-protection system prior to the alarm check valve. The surveyor must verify that there is no connection to the potable water system downstream of the containment backflow protection.

If during the course of a survey an unprotected cross-connection that presents a current risk to public health is identified, and appropriate backflow protection cannot be installed, immediate corrective action must be taken, such as

- Informing water supplier management staff
- Notification of the local health agency or code enforcement official
- "Do not drink" advisory posted at all water outlets
- Termination of water service
- Provision for temporary potable water to occupants

Tracking changes in water use and new customers is a critical part of the survey program. The water supplier should make every attempt to prevent or eliminate cross-connections at the time that water service is established. The water supplier should identify and work closely with local building/plumbing officials, fire inspectors, and health agencies to better accomplish the goal of controlling, protecting, or eliminating cross-connections.

Notification and Enforcement

Following a survey, the surveyor or water supplier must notify the owner or owner's agent in writing of any action required for compliance. Specific information on identified cross-connections, the required or recommended corrective action, along with a reasonable compliance period, should be clearly stated in the initial compliance notification. The timeframe to complete any necessary corrective action should not be arbitrary but based on requirements contained in the local ordinance, and/or on the degree of risk posed by the cross-connections found during the survey. In setting compliance deadlines, consideration should also be given to the complexity and cost of the necessary corrective actions. Dates for compliance established by the water supplier must be tracked and efficiently enforced by the supplier.

Local customer service practices often provide for compliance extensions. Extensions should be granted only when there is no immediate risk of backflow, and the customer has signed an agreement to install the required backflow protection by the extended due date. Obtaining a commitment from the customer creates a compliance partnership, significantly improving the likelihood that the appropriate backflow protection will be installed. This commitment is a compliance plan used to document expectations of the customer and potential actions that the water supplier will take, should formal enforcement be required to accomplish compliance.

To protect public health, customers found to be in violation of the cross-connection control rules should be brought into compliance within a timely manner. The water supplier's ordinance must have specific enforcement options available. These typically include reasonable access to verify compliance, discontinue of water service, fines or assessments, work orders, and in extreme cases, loss of privilege to receive water service. Enforcement action should be progressive in nature and applied in a fair and consistent manner. Unprotected cross-connections that pose an imminent and extreme hazard may require immediate disconnection for public health protection. Reconnection should be allowed only after proper backflow protection is installed.

If enforcement action requires the water supplier to disconnect service, an established list of local agencies must be notified. These may include but are not limited to

- Local health department
- Fire department or fire marshal
- Local law enforcement
- Plumbing code officials
- Appointed or elected officials

Record Keeping

A reliable and efficient system for record keeping is essential to a cross-connection control program and should be observed by the water supplier and cross-connection control professionals. Accurate and defensible records are required by many state and provincial/territorial regulations. The length of time records must be retained by statute varies by jurisdiction. In the absence of specified records retention requirements, seven years is a conservative practice.

The various record-keeping systems utilized range from simple index cards for small water systems to complex electronic databases used by metropolitan water systems. In determining the type of system to deploy, consideration should be given to efficient record searches, and ease of reporting and updating. Customized computer programs provide additional capabilities that may include but are not limited to

- Customized compliance tools and reports
- Water supplier billing system data link
- Paperless communication (text, email, website)
- Automatic letter scheduling and generation
- Mobile workforce integration (tablets and smartphones)
- Access to third party mail services for mass mailing

Regardless of the method used for record keeping, a survey report must be generated to catalog information concerning the facility and the survey findings. A survey report may consist of a standardized form with check boxes, or a unique, comprehensive narrative. Useful information to include in a survey report includes but is not limited to

- Methodology used to conduct the survey
- General facility overview
    - physical address
    - primary contact information
    - water service size(s)
    - meter number(s)
    - facility use
    - photos or sketches
- List of all existing backflow prevention devices and assemblies
    - type
    - size
    - manufacturer
    - model
    - serial number
    - location description/remarks
- Drawings of the facility's potable water piping system
- Identified cross-connections and the associated degree of hazard
- Backflow protection installed (if any)
- Required backflow protection

To streamline the record-keeping process, standardized forms and letters should be developed, including but not limited to

- Survey forms
- Survey and/or assembly testing notification letters
- Noncompliance letters
- Water service termination notice
- Hydrant use authorization forms

Example notices and letters are located for reference in appendix A. Modifications will be required to ensure compliance with regulatory and local program requirements. When communicating with customers, technical jargon and abbreviations should be avoided. References to required code or regulatory compliance should be as specific as possible. Informational brochures or frequently asked question documents should be included with letters to customers.

Cross-connection surveys are conducted to identify and control cross-connections as part of a multiple barrier approach to protecting the drinking water supply, a responsibility that is primarily vested in the water supplier. It is not the responsibility of the water supplier to design the corrective action needed to mitigate the cross-connections, but to assess that the cross-connections are protected or eliminated in conformance with the water supplier's cross-connection control program and/or the plumbing code.

REFERENCE

US Environmental Protection Agency (USEPA). 1999. Guidance Manual for Conducting Sanitary Surveys of Public Water Systems: Surface Water and Groundwater Under the Direct Influence (GWUDI). EPA 815-R-99-016. Washington D.C.: USEPA.

AWWA MANUAL

M14

Chapter 6

Sample Hazards and Proper Protection

The water supplier has the ultimate responsibility for the quality of drinking water in their distribution system. The water supplier can monitor its source and provide treatment to ensure safe drinking water is produced, but it must also ensure that the water users do not affect water quality through unprotected cross-connections that could allow a backflow event to alter water quality. The water supplier has the responsibility to conduct a hazard assessment of each water user to determine the degree of hazard that exists at each property supplied with drinking water. However, for a variety of reasons ranging from legal to logistics, some water suppliers may be unable to conduct a hazard assessment of each water user, making a decision on the proper protection needed a difficult one. Once the water supplier has evaluated all of their water users, they should arrange water users from highest to lowest degree of hazard to prioritize the installation of a backflow preventer where needed.

When changes take place at a water user's facility, such as occupancy change, piping modifications, equipment installation, or other water use changes, the hazard assessment for that property should be revisited. Ideally, the water supplier would reassess or confirm the hazard assessment of all water users at some reasonable frequency. The ability to conduct hazard assessments is impacted by the level of staffing. Small water systems are especially challenged with regard to resources, but these assessments cannot be ignored. Once the hazard assessment for a water user is completed, the appropriate type of backflow protection can be established. If the water supplier can rely on the property owner to install and maintain proper backflow protection at each existing hazard within the property and if the property owner does not make modifications to the water usage without installing the proper backflow protection, the level of backflow protection needed at the service connection may be reduced.

Some water suppliers may only be able to assign a degree of hazard based on an off-site evaluation of the type of water service connection or account for a specific property (i.e., commercial, industrial, residential, etc.). This should not be relied on and a hazard assessment should be conducted when possible. As mentioned in chapter 2, the terminology used to describe the degrees of hazard vary; however, in this chapter the terms *high* and *low* will be used.

It is worth reiterating the purpose of this manual from chapter 1: "This manual provides guidance to water suppliers on the recommended procedures and practices for the operation of a cross-connection control program." With that in mind, this chapter will discuss the hazards that exist within various types of properties, businesses, and facilities. The goal of the discussion is to assist the water supplier in determining the risks that exist should backflow occur from a specific service connection.

The water supplier should first evaluate the hazards that exist at their own facilities such as office buildings, pumping plants, storage facilities, and treatment plants. When a hazard is identified, the water supplier is obligated to install and maintain the appropriate backflow preventers to protect the quality of the water within the distribution system, thereby ensuring the delivery of safe drinking water to the very first customers—its staff.

TYPICAL HAZARDS

Auxiliary Water

The term *auxiliary water* (defined in the glossary) is commonly used to describe water supplies or sources that are not under the direct control or supervision of the water supplier. These water supplies are typically natural waters derived from wells, springs, streams, rivers, lakes, harbors, bays, and oceans, including water delivered to premises from another water supplier. Used waters that have passed beyond the water supplier's control at the point of delivery and that may be stored, transmitted, or used in a way that may have contaminated them are considered auxiliary water supplies. With the growing demand to conserve potable water, the treatment and distribution of used waters, including wastewater, is increasing and requires particular attention from the water supplier.

Some typical used water supplies include

- water in industrialized water systems
- water in reservoirs or tanks used for fire-fighting purposes
- irrigation reservoirs
- swimming pools, fish ponds, and mirror pools
- memorial and decorative fountains and cascades
- cooling towers
- baptismal, quenching, washing, rinsing, and dipping tanks
- reclaimed water and recycled water
- gray water and spray aerobic water

All of these supplies pose a potential hazard to the quality of the water in the water supplier's distribution system. These waters may become contaminated because of industrial processes; through contact with the human body, dust, vermin, birds, etc.; or by means of chemicals or organic compounds that may be introduced into tanks, lines, or systems for control of scale, corrosion, algae, bacteria, odor, or for similar treatment.

In evaluating the hazard an auxiliary water supply poses to the water supplier's distribution system, it is not necessary to determine that the auxiliary water sources are developed and connected to the potable water system through cross-connections. It is only necessary to determine that the nonpotable sources are available to the premises and are of a quantity sufficient to make it desirable and feasible for the customer to develop and use.

Commercial, Industrial, and Institutional Facilities

Examples of these types of facilities are: beverage bottling plants, breweries, canneries, packing houses, food service restaurants, reduction plants, chemical plants, manufacturing plants, dairies, cold storage plants, film laboratories, hospitals, medical offices and facilities, clinics, morgues, mortuaries, animal clinics, commercial laundries, marine facilities, multistory buildings, oil and gas production, paper and pulp plants, plating facilities, radioactive material-handling facilities, wastewater treatment plants, water treatment plants, daycares, long-term care facilities, retail stores, professional offices, and many other commercial businesses.

The hazards normally found in these facilities are carried by the water coming through cross-connections between the potable water system and steam-connected facilities, such as pressure cookers, autoclaves, washers, cookers, tanks, lines, flumes, and other equipment used for storing, washing, cleaning, blanching, cooking, flushing, or fluming, or for transmission of foods, fertilizers, wastes, can- and bottle-washing machines, lines where caustics, acids, detergents, and other compounds are used in cleaning, sterilizing, flushing; controlling scale in cooling towers, and circulating systems that may be contaminated with algae, bacterial slimes, toxic water-treatment compounds, industrial fluid such as cutting and hydraulic fluids, coolants, hydrocarbon products, glycerin, paraffin, caustic and acid solutions; water-cooled equipment that may be sewer-connected, such as compressors, heat exchangers, air-conditioning equipment; fire-fighting systems that may be subject to contamination with antifreeze solutions, liquid foam concentrates, or other chemicals or chemical compounds used in fighting fire; fire systems that are subject to contamination from auxiliary or used water supplies; or industrial fluids.

Fire Hydrants

Fire hydrants are installed primarily to provide a water supply for firefighting purposes. However, fire hydrants also provide access for contaminants to enter the water distribution system. To ensure public health and safety, fire hydrants supplied from potable water systems should be monitored regularly and maintained as required by the water supplier and the fire authority. The water supplier should also consider fire-fighting equipment and the use of chemicals with that equipment to ensure the potable water system is not contaminated.

Fire hydrants are used for purposes other than firefighting, e.g., for construction water, dust control, water hauling, jumper connections for superchlorination of mains, pressure testing, and temporary service. These uses present the potential for many different types of cross-connections and the introduction of contaminants.

Protection recommended. The water supplier should develop a standard operating procedures (SOP) to indicate when to provide contractors, businesses, or residents with a hydrant meter and backflow device if they have a legitimate need to access bulk water from a fire hydrant. The SOP establishes guidelines and policies for financial accountability of water usage as well as continued protection from unknown materials, chemicals, and other nonpotable substances that are contained in or that have been in contact with hoses, pipes, tanks, etc.

The hydrant meter is a portable water meter used to measure the amount of water that flows through the fire hydrant for bulk water use. If there is any question regarding the degree of hazard, a reduced-pressure principle backflow prevention assembly is recommended. On mobile tanks, an air gap is a beneficial additional safeguard. However, because the air gap does not ensure the integrity of the materials supplying the tank, it should be accepted as a sole means of backflow protection only if the water supplier can guarantee that the piping or other conduit to the air gap will remain in a potable state at all times.

Fire-Sprinkler and Water-Based Fire-Protection Systems

There are a variety of commercial water-based fire protection systems including fire-sprinkler systems, standpipe systems, and foam/water systems.

The water supplier must be mindful of not only state and provincial regulations pertaining to fire-sprinkler systems but also local regulations. These regulations may limit the water supplier's options for requiring backflow preventers on new and existing fire service lines. The water supplier also must understand that installing a backflow preventer on an existing fire-suppression system may have a significant adverse effect on the hydraulic performance of the system. This is especially relevant when the original design may not have included a backflow preventer. If there is an existing backflow preventer that needs to be replaced on an existing system, the replacement backflow preventer must be of the same size and type (e.g., 4-in. double check assembly) or calculations must be done to show that the water-based fire-protection system will still function as it was designed.

Backflow prevention devices such as double check valve assemblies, reduced pressure principle backflow prevention assemblies, and the detector versions of these assemblies are generally designed with internally loaded check valves that are continuously pressed in the closed position. When these devices are used on plumbing and mechanical systems with regular flow, the internal springs are worked and the check valves are released from their rubber seats. Water-based fire-protection systems do not experience regular flow with the frequency of plumbing or mechanical systems. As such, when backflow prevention devices are installed in water-based fire protection systems, a mechanism for creating forward flow needs to be installed downstream so that the internal springs in the backflow device can be exercised and the clapper can be moved off of its rubber seat on a periodic basis. This mechanism for creating flow downstream of the backflow preventer must be capable of meeting the flow demand of the largest single fire-protection system downstream of the backflow device.

Two approaches may be used to assess the water supplier's hazard from its customers' fire-protection systems. The first approach is to consider all types of fire-protection systems (e.g., wet pipe, dry pipe) as hazards that require either a reduced-pressure principle backflow prevention assembly or a double check valve assembly. The second approach is to make a detailed assessment of each type of fire-suppression system.

If the first approach is followed, the water supplier recognizes that any fire-protection system may be operated in a manner other than the manner for which it was originally designed. For example, a dry-pipe system may inappropriately be operated as a wet system during most of the year and then be charged with air for the months of freezing weather. Or, a maintenance contractor could inappropriately add antifreeze to a wet system that was designed to operate without added chemicals. (NOTE: Antifreeze systems that are connected to a potable water supply are required to be food grade. See National Fire Protection Association [NFPA] Standard 13, *Installation of Sprinkler Systems*.) This approach, while it sounds attractive due to its simplicity, may over simplify the problem causing unnecessary expenses for the water supplier's customers.

Protection recommended. For the first approach, a reduced-pressure principle backflow prevention assembly is recommended if there is a high hazard (e.g., risk of chemical addition or alternate water source). A double check valve assembly should be used on all other systems constructed of nonpotable water pipe.

If the second approach is followed, subsequent sections of this chapter provide guidance for the following types of fire-suppression systems:

- Dry-pipe pre-action and deluge fire-suppression systems
- New wet-pipe fire-protection systems
- Existing wet-pipe fire-protection systems

Fire-Protection Systems, Dry-Pipe, Preaction, and Deluge

Dry-pipe systems typically are charged with air or nitrogen. Sprinklers or other outlets are closed until there is sufficient heat to open a sprinkler; then air is released and followed by water, which suppresses the fire. Preaction systems contain air that may or may not be under pressure. They use automatic sprinklers and have a supplemental fire-detection system installed in the same areas as the sprinklers. Deluge systems are open to the atmosphere and the sprinklers or other outlets are open, ready to flow water at all times. When these fire-protection systems are directly connected to a public potable water main line, they do not present a health hazard to the public water system unless chemicals are added to the water as it enters the system.

Protection recommended. A reduced-pressure principle backflow prevention assembly is recommended where there is a high hazard (e.g., risk of chemical addition). A double check valve assembly should be used on all other systems.

Fire-Protection Systems, Wet Pipe

Wet-pipe fire-sprinkler systems contain water that is connected to the potable water system. It has been shown that water contained in closed or nonflow–through fire systems may be stagnant or contaminated beyond acceptable drinking water standards. Some contaminants found in fire-sprinkler systems are antifreeze, chemicals used for corrosion control or as wetting agents, oil, lead, cadmium, and iron. Requirements stipulated by the latest model building/plumbing codes and Occupational Safety and Health Administration (OSHA) regulations require the installation of an approved backflow assembly for all new wet-pipe sprinkler systems.

Protection recommended. For existing wet-pipe fire sprinkler systems that pose only a low-hazard threat, the water supplier may consider an alternative to installing an approved backflow prevention assembly. For new systems, a reduced-pressure principle backflow prevention assembly, reduced-pressure principle detector backflow prevention assembly, or an air gap is recommended if there is a high hazard. A double check valve assembly, double check detector backflow prevention assembly, or air gap is recommended for all other closed or nonflow–through systems.

For existing systems, an air gap, reduced-pressure principle backflow prevention assembly, or reduced-pressure principle detector backflow prevention assembly is recommended where there is a high hazard.

For systems presenting a low hazard only and having a modern Underwriters' Laboratories (UL)-listed alarm check valve that contains no lead, the check valve should be maintained in accordance with NFPA Standard 25, *Standard for the Inspection, Testing, and Maintenance of Water-Based Fire Protection Systems*. When an existing sprinkler system is expanded or modified, requiring a hydraulic analysis, a double check valve assembly should be installed.

For existing systems that present a low hazard and that have an alarm check valve containing lead, a UL-classified double check valve assembly should be installed. See NFPA Standard-13 for information about alarm provisions.

Notes:
- Before installing or testing a backflow prevention assembly on a fire-sprinkler system, the fire authority that has jurisdiction should be consulted for additional criteria that may be required. Additionally, a thorough hydraulic analysis should be performed before installing a backflow prevention assembly on a fire system.
- Water suppliers should consider allowing an existing fire-sprinkler system already equipped with a double check valve assembly on the service connection to install a reduced-pressure principle backflow prevention assembly on an added antifreeze loop(s).
- Water suppliers should evaluate the presence, use, and potential hazard of fire department connections to determine whether a fire system should be classified as a high hazard because of chemicals or auxiliary water sources used by the fire authority.

Fire-Sprinkler Systems, One- and Two-Family Residential

Residential fire-sprinkler system requirements are addressed in NFPA Standard-13. These systems include stand-alone, passive purge, and multipurpose fire-sprinkler systems.

Protection recommended. Multipurpose and passive purge fire-sprinkler systems require no additional backflow prevention protection besides the protection required for the domestic water service. Residential fire-sprinkler systems that are constructed of materials approved for potable water and are flow-through (not closed) systems do not require the installation of a backflow assembly. The ends of these systems are connected to a fixture that is regularly used. This prevents the water in the system from becoming stagnant.

The protection recommended for stand-alone fire-sprinkler systems in residential installations depends on the system hazard. This hazard is determined by chemicals or antifreeze additives, piping or system materials, or a number of other factors. A reduced-pressure backflow prevention assembly or a reduced-pressure detector assembly is recommended for systems deemed a high hazard. A double check valve assembly or double check detector assembly is recommended for systems deemed low hazard.

Irrigation Systems

Irrigation systems include but are not limited to agricultural, residential, and commercial applications. They may be connected directly to the public water system, internally to the private plumbing system, or to both systems. However, the irrigation system is a high hazard for several reasons. Most systems are constructed of materials that are not suitable for use with potable water. Sprinklers, bubbler outlets, emitters, and other equipment are exposed to substances such as fecal material, fertilizers, pesticides, and other chemical and biological contaminants. Sprinklers generally remain submerged in water after system use or storms. Irrigation systems can have various design and operation configurations. They may be subject to various onsite conditions such as additional water supplies, chemical injection, booster pumps, and elevation changes. All of these conditions must be considered in determining proper backflow protection.

Solutions of chemicals and/or fertilizers are used in or around irrigation systems for many purposes. Some of the chemical compounds that may be injected or aspirated into or come in contact with these systems include

- Fertilizers: Ammonium salts, ammonia gas, phosphates, potassium salts

- Herbicides: 2,4-D, dinitrophenol, 2,4,5-T, T-pentachlorophenol, sodium chlorate, borax, sodium arsenate, methyl bromide
- Pesticides: TDE, BHC, lindane, TEPP, parathion, malathion, nicotine, MH, and others
- Fecal matter: Animal and other

Protection recommended. Irrigation systems connected to the public water system should be considered a high hazard, and the appropriate protection is an air-gap separation or a reduced-pressure principle backflow prevention assembly.

A properly installed pressure vacuum breaker assembly (PVB) may be used for service protection if the water service is a dedicated supply to the premises or property and used strictly for irrigation (such as for median islands and parking strips). The installation requirements for a PVB must be closely adhered to, requiring proper elevation and no means or potential means for backpressure. The local authority having jurisdiction should be consulted.

Irrigation systems connected internally to the private plumbing system are considered a high hazard. Appropriate backflow protection for such internal applications is outlined in the local plumbing code. However, if the system is not protected according to the water supplier's requirements or if the water supplier lacks jurisdiction over internal plumbing code compliance, protection should be required at the service connection as previously outlined.

Marine Facilities and Dockside Watering Points

The actual or potential hazards to the potable water system created by any marine facility or dockside watering point must be individually evaluated. The basic risk to a potable water system is due to the possibility that contaminated water can be pumped into the potable water system by the fire pumps or other pumps aboard ships. In addition to the normal risks peculiar to dockside watering points, risks are found at those areas where dockside watering facilities are used in connection with marine construction, maintenance and repair, and permanent or semipermanent moorages. Health authorities point out the additional risk of dockside water facilities that are located on freshwater or diluted salt water where, if backflow occurs, it can be more easily ingested because of the lack of salty taste.

Protection recommended. Minimum system protection for marine installations may be accomplished in one of the following ways:

- Where water is delivered directly to vessels for any purpose, a reduced-pressure principle backflow prevention assembly or air gap must be installed at the pier hydrants. All dockside hydrants that are used (or are available for use) to provide water to vessels should be so protected. If an auxiliary water supply such as a saltwater fire system is used, the entire dockside area should be isolated from the water supplier's system by an approved air gap. Where water is delivered to marine facilities for fire protection only and no auxiliary supply is present, all service connections should be protected by a reduced-pressure principle backflow prevention assembly. If hydrants are available for connection to a vessel's fire system, a reduced-pressure principle backflow prevention assembly should be installed at the user connections as well.

- Where water is delivered to a marine repair facility, a reduced-pressure principle backflow prevention assembly should be installed at the user connection. Where water is delivered to small-boat moorages that maintain hose bibs on a dock or float, a reduced-pressure principle backflow prevention assembly should be installed at the user connection and a hose connection vacuum breaker should be installed on each hose bib. If a sewage pump station is provided, the area should be isolated by installation of a reduced-pressure principle backflow prevention assembly. Water used for fire protection aboard ships connected to dockside fire hydrants should not be taken aboard from fire hydrants unless the hydrants are on a fire system that is separated from the domestic system by an approved reduced-pressure principle backflow prevention assembly or unless the hydrants are protected by portable, approved reduced-pressure principle backflow prevention assemblies.

Reclaimed or Recycled Water

Generally, reclaimed water is treated sewage effluent that undergoes further treatment to improve its quality. It is critical to remember that reclaimed water is nonpotable. Reclaimed water use is not new; some use dates back to the 1920s. Currently, reclaimed water has become a popular means to conserve potable water. More extensive reclaimed-water distribution systems are being constructed and expanded. Reclaimed water is regularly used for irrigation applications, including golf courses, parks, ball fields, schools, median islands, cemeteries, and commercial and residential landscapes. In some areas, reclaimed water is used for other applications, such as cooling towers, toilet flushing, dust control, and general construction purposes.

Although reclaimed water systems are a tremendous conservation resource, they do pose new risks if they are improperly maintained, operated, or identified. To ensure public health and safety, safeguards must be put in place and maintained at all times. Safeguards include the installation of proper backflow prevention on the potable water systems and the assurance that there is not a direct cross-connection between a site's potable water system and the reclaimed water system. Several test methods are used to ensure against existing cross-connections. All involve pressure testing in one form or another and, in some instances, they involve using a dye that is safe for potable water. When selecting or developing a test method to ensure system separation, potable water should be used initially to test the separation of the proposed reclaimed system. This precaution prevents accidental contamination of the potable water system if an unknown cross-connection exists between the proposed new or a converted existing system to be supplied by reclaimed water and the site's potable water system.

Other precautions. Water suppliers that supply reclaimed water to customers should conduct thorough site evaluations before startup and periodically thereafter. There should be no locations on the site where ponding could occur or where reclaimed water overspray could contact any human, food, or drinking areas. Signs specifying the use of reclaimed water should be posted at all points of entry to the site. Standard hose bibs should not be used on the reclaimed water system. All reclaimed water piping, valves, meters, controls, and equipment should be clearly labeled and marked. Reclaimed water lines should be separated from potable water lines following the same requirements for the separation of potable water lines from sewage lines.

Protection recommended. An air-gap separation or a reduced-pressure principle backflow prevention assembly is recommended on each potable water line entering a reclaimed water use site.

Note: Where reclaimed water is used for an industrial purpose, such as cooling towers, a backflow preventer may be on the reclaimed-water line to such equipment to prevent the backflow of chemicals that are added to the equipment for maintaining pH levels, corrosion control, etc., from entering the reclaimed water system. Field-test equipment used in testing these backflow prevention assemblies on reclaimed water should be clearly identified as nonpotable field-test equipment and should be kept separate from field-test equipment used to test backflow prevention assemblies supplied by potable water.

Residential Water Services

Residential water services are a single water service providing water to structures providing a living place for a single individual, couple, or single family. These water services generally present a low hazard to the water supplier's distribution system.

Some items to consider in establishing a degree of hazard and in requiring backflow protection on residential water services are pets, livestock, fish, chemicals, pools, fountains, tanks, irrigation, dialysis equipment, photo developing equipment, gray water, reclaimed water, an auxiliary water supply, heating and cooling equipment, and other equipment or operations that use water. The elevation of the site's plumbing system above the water service connection also should be considered. If the site does not have one of the aforementioned hazards and the plumbing system meets and is maintained to current plumbing code requirements, the water supplier may elect to forego service protection.

Typically, residential plumbing systems are smaller than most commercial plumbing systems. Therefore, the adverse effects of thermal expansion can be greater and more immediate. As required with any closed system, thermal-expansion protection that meets the plumbing code must be installed and maintained to ensure the safety and longevity of the private plumbing system.

The water supplier should actively work to educate residential customers, as well as the entire community, regarding the hazards of backflow.

Restricted, Classified, or Other Closed Facilities

A service connection to any facility that is not readily accessible for inspection by the water supplier because of military or industrial secrecy requirements or other prohibitions or restrictions should be categorized as posing a high hazard. In selecting the protection recommended, the potential for cross-connection to sewer systems should be considered.

Solar Domestic Hot-Water Systems

The hazards normally found in solar domestic hot-water systems include cross-connections between the potable water system and heat exchangers, tanks, and circulating pumps. Depending on the system's design, the heat transfer medium may vary from domestic water to antifreeze solutions, corrosion inhibitors, or gases. The degree of hazard will range from a low hazard when potable water is used to a high hazard when a toxic transfer medium, such as ethylene glycol, is used. Contamination can occur when the piping or tank walls of the heat exchanger between the potable hot water and the transfer medium begin to leak.

Liquid-to-liquid solar heat exchangers can be classified as follows:

- *Single wall with no leak protection (SW)*: A heat exchanger that provides single-wall separation between the domestic hot water and the transfer medium. Failure of this wall will result in a cross-connection between the domestic hot water and the heat transfer medium.

- *Double wall with no leak protection (DW)*: A heat exchanger that has two separate, distinct walls separating the potable water and the transfer medium. A cross-connection between the potable hot-water system and the transfer medium requires independent failure of both walls.

- *Double wall with leak protection (DWP)*: A heat exchanger that has two separate, distinct walls separating the potable hot water and the transfer medium. If a leak occurs in one or both walls of the DWP, the transfer medium will flow to the outside of the heat exchanger, thus indicating the leak.

Table 6-1   Recommended protection for solar domestic hot-water systems

Hazard Rating of Transfer Medium	Heat Exchanger	Protection Recommended
Nonhealth	SW	DC
Nonhealth	DW, DWP	None*
Health	SW, DW	RP
Health	DWP	None

* Some jurisdictions may require backflow protection and/or require all heat exchangers to be DWP. Check local plumbing codes.

Note: DC = double check valve assembly; DW = double wall with no leak detection; DWP = double wall with leak detection; RP = reduced-pressure principle backflow prevention assembly; SW = single wall with no leak detection.

Protection recommended. The recommendations in Table 6-1 are to be used as a guide to recommend protection for solar domestic hot-water systems.

Water-Hauling Equipment

This category includes any portable or nonportable spraying or cleaning units that can be connected to any potable water supply that does not contain proper backflow protection.

The hazards normally found with water-hauling equipment include cross-connections between the potable water system and tanks contaminated with toxic chemical compounds used in spraying fertilizers, herbicides, and pesticides; water-hauling tanker trucks used in dust control or for domestic wells or cisterns; and other tanks on cleaning equipment.

Protection recommended. An air-gap separation or a reduced-pressure principle backflow prevention assembly installed at the point of the connection supplying such equipment is recommended. Hoses or piping to the equipment may not be of potable quality or may have been in contact with contaminants. In all cases, the regular inspection, testing, and maintenance of backflow prevention assemblies on portable units is essential. A water supplier may wish to designate specific watering points, such as those equipped with air gaps, for filling portable units. This provides better monitoring abilities for the water supplier.

HAZARDS POSED BY A WATER SUPPLIER

Distribution System

In some instances, the water supplier may have made connections to the distribution system that pose a risk. These instances fall into three different types of cross-connections. The first type is used for draining tanks, reservoirs, and mains; to facilitate air release and vacuum relief in mains; for fire hydrants and other appurtenances with underground drain ports; for irrigation-system connections at reservoirs, wells, and booster sites; and for potable backup supply to reclaimed water reservoirs. The second type includes temporary connections used for the direct supply of water for construction and maintenance work (e.g., disinfection of new mains); the filling of tanker trucks or trailers, or water haulers (e.g., construction water for dust control and water for pesticide applications or firefighting, delivery of potable water to a residential holding tank); and flushing of sewers and storm drains. The third type includes inadvertent connections. For example, there have been reports of systems inadvertently connecting sewer lines directly to potable water lines after repairs.

Protection recommended. For permanent connections
- Reservoirs and storage tanks: A screened air-gap separation is recommended on overflow pipes.
- Air-release and vacuum valves: A screened air-gap separation is recommended on air-discharge outlet pipes.
- Fire hydrants and other appurtenances with underground drain ports: Eliminate all underground drain connections wherever possible. For dry=barrel fire hydrants, no recommended protection is presently available.
- Irrigation-system connections: Onsite irrigation systems are considered a high hazard and should have backflow protection installed according to the plumbing code.
- Backup supplies to reclaimed-water reservoirs and tanks: This is a high hazard and a screened air gap installed on the inlet pipe to the vessel is recommended.

For temporary connections:
- Supply of water for filling or disinfecting new mains, etc.: A reduced-pressure principle backflow prevention assembly is recommended.
- Supply of water for construction sites, filling tanks, etc.: An air-gap separation or a reduced-pressure backflow prevention assembly is recommended.
- Supply of water for sewer flushing: Because of the high hazard and the difficulty in monitoring the maintenance of an air gap (or, where allowed, a reduced-pressure backflow prevention assembly), the use of water from hydrants is prohibited for sewer flushing and all sewer flushing water should be provided from tanker trucks.

NOTE: On mobile tanks, an air gap remains a good additional safeguard. However, because it does not ensure the integrity of the materials supplying the tank, an air gap should be accepted as a sole means of backflow protection only if the water supplier is certain that the piping or other conduit to the air gap will remain in a potable state at all times.

Water Treatment Plants

Water treatment plants have many situations and equipment that pose a potential or actual threat to the quality of water delivered to its customers. Water treatment plant cross-connection control should provide protection of water-using equipment, quality assurance for the finished water, and protection for the distribution system starting with the first customer—the water treatment plant staff.

Hazards may exist where
- There is treatment for surface water or groundwater under the direct influence of surface water where the primary hazard is microbiological. The backflow of raw water into the potable water system could be a high hazard.
- During the treatment process for the removal of chemical contaminants, a high hazard would exist if the raw water flowed back into the potable water.
- There is treatment for a secondary chemical contaminants (e.g., removal of manganese or iron from groundwater) that would constitute a low hazard if the raw water flowed back into the potable water.
- During the treatment process, there are waste products that must be discharged. It could be discharged to a holding pond or a reclaimed water tank for reprocessing, to a backwash discharge swale or pond, or to a sewer. Proper separation between finished water and all discharged water must be maintained.

- Laboratory facilities are operated for quality control and compliance. Even simple aeration and pressure filtration systems for manganese removal would use a field-test kit containing cyanide. Care must be taken with all chemicals used at water treatment plants.
- Another hazard in water treatment plants is the treatment chemical supply. Backflow could potentially occur that could introduce the chemicals at higher concentrations than intended, such as chlorine solution.

Protection recommended. For the protection of the water supplier's employees, the water supplier shall comply with the plumbing and safety codes that normally govern private property, even if the water supplier's plant is exempt from normal plumbing inspection requirements. All pipes in the treatment system must be labeled, from the raw water source to the finished water pipe leaving the plant. It must be clear which pipes are carrying potable water, nonpotable water, or chemicals.

The facility will contain a limited number of dedicated and labeled outlets for potable water in the treatment plant. These potable connections should be regularly evaluated to ensure no cross-connection has been created. The facility should provide the normal plumbing fixtures for building occupancy (e.g., toilets and sinks) in a separate building or in a building addition to the treatment plant and provide proper backflow protection for the potable water line into the plant and laboratory facilities.

To maintain water quality, the water supplier should use proper backflow prevention assemblies to protect the integrity of each process in the treatment plant. This should be done even though the water piping may be labeled nonpotable and thus may not be subject to plumbing code fixture-protection requirements (see Table 6-2).

Table 6-2 Recommended protection at fixtures and equipment found in water treatment plants

Description of Fixture, Equipment, or Use	Recommended Minimum Protection
Raw water storage reservoir	Screened air gap on overflow Air gap on drain No bypass for high hazard
Bulk chemical storage	Air gap on dilution water supply Provide day tank with screened air gap on overflow and air gap on drain
Filter bed/filter tank discharge	Air gap on waste discharge
Surface washer	Pressure vacuum breaker assembly/double check valve assembly
Chemical-feed pumps	Ensure discharge at point of positive pressure, and antisiphon valve Check valve at discharge point Foot valve in tank No pump primer line
Chemical-feed injectors	Check valve at discharge point Check valve at injector inlet
Saturators and dry-chemical solution tanks	Air gap on fill line Screened air gap on overflow Air gap on drain
Membrane clean-in-place systems	Provide physical disconnect
Sample lines to monitoring equipment	Air gap or atmospheric vacuum breaker Label "nonpotable water"
Hose-bib connections	Hose-bib vacuum breaker

SAMPLE HAZARDS AND PROPER PROTECTION   87

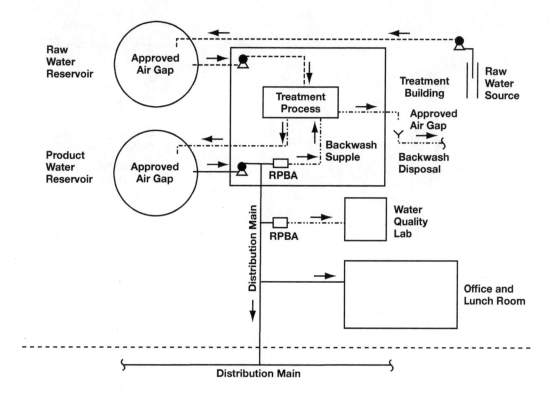

Figure 6-1   Cross-connection control, water treatment plants

For internal protection and quality assurance, the water supplier should

- use approved air gaps or testable backflow prevention assemblies wherever possible, periodically inspecting air gaps and testing and maintaining backflow prevention assemblies.

- provide proper backflow protection based on the degree of hazard.

Containment or premises isolation may not be practical in normal terms. The in-plant potable water may be supplied from the treatment system's finished water line. This could be the case for small, pressurized filter systems that send water into the distribution main for transmission to a remote off-site reservoir.

Containment is applicable where the treatment system sends water into a finished water storage tank for pumping through a transmission line to a remote reservoir. Pressurized water may be brought into the plant from the distribution main.

Figure 6-2 shows the general arrangement of service-containment and area-isolation backflow prevention. The water supplier should apply to the water treatment plant the same containment requirements imposed on industrial customers. Table 6-3 lists the backflow-prevention measures used to prevent cross-connections that could occur between individual treatment processes within the treatment plant from contaminating the finished product.

Supplier-Owned Offices and Work Areas

Supplier-owned offices and work areas contain the same cross-connection hazards found on customers' premises. The water supplier should follow the same cross-connection control requirements it imposes on its customers.

AWWA Manual M14

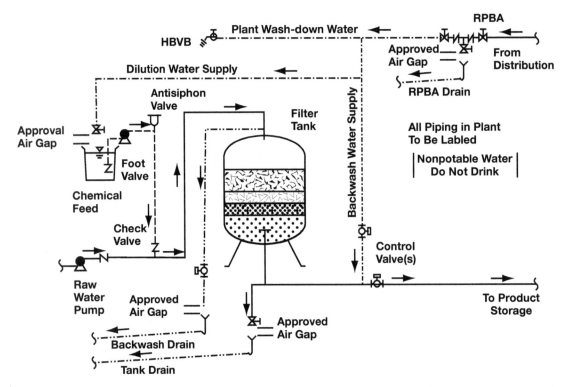

Figure 6-2 Service-containment and area-isolation water treatment plants

Protection recommended. Service protection by a containment backflow preventer following the policy established by the water supplier for customers is recommended. In addition, compliance with the fixture-protection backflow requirements established by the plumbing code having jurisdiction is recommended.

There are two locations for the installation of backflow prevention assemblies: containment or service protection and isolation or internal protection. Containment or service protection is the most common approach taken by water suppliers as it helps ensure that no contaminant can backflow into the distribution system. Table 6-3 can be used for quick reference. However, information should be provided for determining the appropriate degree of hazard and corresponding protection for a specific piece of equipment or situation within a building/facility or a water treatment plant (isolation/internal protection). In fact, as mentioned throughout this chapter, the water supplier needs to inspect its own facilities starting with the public water system itself to assess the degree of hazard posed by various pieces of equipment and install and maintain the appropriate backflow preventers.

The responsibility to evaluate internal hazards and to require the separation of potable water piping from piping, equipment, fixtures, operations, or other water uses that might pollute or contaminate the potable water (i.e., isolation or internal protection) generally rests with the authority having jurisdiction. This chapter's discussion of the typical hazards found within the customer's premises is provided only for the water supplier's overall assessment of the customer's plumbing system. It is not intended to provide guidance regarding internal protection for compliance with the plumbing code or with other regulations pertaining to the customer's property.

PROTECTION FOR SPECIFIC CUSTOMERS

Tables 6-3 through 6-6 provide information on protection for specific customers.

Table 6-3 For service protection (containment)

Type of Service Connection	Hazard	Backflow	AG*	RP	DC	RPDA	DCDA
Commercial	High†	BP		x			
		BS	x	x			
	Low‡	BP		x	x		
		BS	x	x	x		
Fire protection	High	BP		x		x	
		BS	x	x		x	
	Low	BP		x	x	x	x
		BS	x	x	x	x	x
Industrial	High	BP		x			
		BS	x	x			
	Low	BP		x	x		
		BS	x	x	x		
Institutional	High	BP		x			
		BS	x	x			
	Low	BP		x	x		
		BS	x	x	x		
Residential	High	BP		x			
		BS	x	x			
	Low	BP		x	x		
		BS	x	x	x		
Premises with an auxiliary source of water	High	BP		x			
		BS	x	x			
	Low	BP		x	x		
		BS	x	x	x		

* The use of an air gap at a service connection is highly unlikely but does provide the most reliable protection and should be considered where a premises is able to develop its own pressurized distribution system or piping configuration.
† High hazard indicates a health hazard.
‡ Low hazard indicates a nonhealth hazard.

NOTE: AG = air gap; RP = reduced-pressure principle; DC = double check; RPDA = reduced-pressure detector assembly; DCDA = dual-check detector assembly.

Courtesy of Nick Azmo

Table 6-4   Containment protection

Backflow Prevention Product	Hydraulic Application	Health Hazard	Primary Domestic Water Applications
Reduced-pressure principal backflow assembly (RP)*	Backpressure and backsiphonage	High[†] and Low[‡]	high or low hazard containment protection
Double check valve assembly (DCVA) *	Backpressure and backsiphonage	Low	low hazard containment protection including fire systems without the addition of chemical additives

* Assemblies should not be installed in meter pits, vaults, or confined spaces. Entry into a confined space should require an OSHA 2264 certification. Assemblies should not be subject to flooding or freezing conditions.
[†] High hazard indicates a health hazard.
[‡] Low hazard indicates a nonhealth hazard.

Courtesy of Nick Azmo

Table 6-5   Typical backflow prevention devices

Backflow Prevention Product	Hydraulics	Hazard	Typical Applications
Atmospheric vacuum breaker*	Backsiphonage	High[†] and Low[‡]	Typical installations include: lavatories, process tanks, sinks, dishwashers, soap dispensers, lawn irrigation systems
Backflow preventer with an intermediate atmospheric vent	Backpressure and backsiphonage	Low	Installed where there is low or atmospheric pressure at the outlet. Typical installations include residential boilers, food steamers, lab equipment, process tanks
Dual-check valve backflow preventer	Backpressure and backsiphonage	Low	Typically installed on individual outlets of residential water services

* Mechanical devices are subject to failure and should be replaced every five years or per the manufacturer's recommendations.
[†] High hazard indicates a health hazard.
[‡] Low hazard indicates a nonhealth hazard.

Table 6-6 Irrigation and hose connection protection

Backflow Prevention Product	Hydraulic Application	Health Hazard	Typical Primary Domestic Water Application
Pressure vacuum breaker assembly*	Backsiphonage	High[†] and Low[‡]	Typical installations include outdoor lawn irrigation systems, process tanks, dishwashers, soap dispensers
Spill resistant vacuum breakers*	Backsiphonage	High and Low	This assembly is suitable for indoor applications. Dental equipment, vats, x-ray equipment
Reduced pressure	Backpressure and Backsiphonage	High and Low	Any hazard application. Required when chemicals are applied
Atmospheric vacuum breaker**	Backsiphonage	High and Low	Typical installations include lavatories, process tanks, sinks, dishwashers, lawn irrigation systems, soap dispensers
Hose connection vacuum breaker**	Backpressure Low head (<10 ft) and Backsiphonage	High and Low	This device is typically installed at the hose bibb connection
Vacuum breaker wall hydrants, freeze resistant, automatic draining type **	Backpressure Low head (<10-Ft) and Backsiphonage	High and Low	This device is typically installed at the hose bibb connection.
Hose connection backflow preventers **	Backpressure Low head (<10-Ft) and Backsiphonage	High and Low	This device is typically installed at hose connections.
Freeze resistant sanitary yard hydrants w/ backflow protection **	Backpressure Low head (<10-Ft) and Backsiphonage	High and Low	This device is typically installed to protect yard hydrants that require freeze protection.

* When any chemical is introduced into the lawn irrigation system, a reduced-pressure backflow assembly should be installed on the system.
[†] High hazard indicates a health hazard.
[‡] Low hazard indicates a nonhealth hazard.
**Mechanical devices are subject to failure and should be replaced every five years, or per the manufacturer's recommendations.

Courtesy of Nick Azmo

REFERENCES

National Fire Protection Association (NFPA). 2013. Standard 13, *Installation of Sprinkler Systems*. Quincy, MA: NFPA.
NFPA. 2014. Standard 25, *Standard for the Inspection, Testing, and Maintenance of Water-Based Fire Protection Systems*. Quincy, MA: NFPA.

This page intentionally blank.

Appendix A

Example Notices and Letters

Example #1: Cross-Connection Control Program Survey Notice

The Purpose of the <Any City USA's> Cross-Connection Control Program, as defined in the local Ordinance, is to help eliminate possible contamination of the public water distribution system. There are two required components of the program: (1) site survey, and (2) testing of backflow prevention assemblies.

The <Any City USA> will be working to conduct these surveys. Thank you in advance for your cooperation in this matter.

As part of this program, a survey of your facility's internal water system is to be completed. Inspectors will be reviewing your water system for connections that could possibly contaminate the water distribution system. The survey is tentatively scheduled for (list date), our inspector will do their best to be on site this day; however, we may be on site a day or two before or after the scheduled date. The survey must be completed during normal business hours 8:00 AM to 5:00 PM. If you need a more specific time please call (phone number) to arrange an appointment.

Any costs associated with the replacement, modification, installation and/or testing of backflow prevention assemblies is the responsibility of the property owner/manager and/or occupant.

You will be notified following the survey if modification(s) and/or testing of backflow prevention assemblies are necessary. We look forward to working with you in protecting the drinking water supply. If you have any questions or concerns, please contact:

_____

Example #2: Cross-Connection Control Program Survey Compliance Notice

The purpose of the <Any City USA's> Cross-Connection Control Program, as defined in Ordinance, is to help eliminate possible contamination of the public water distribution system. There are two required components of the program: (1) site survey, and (2) testing of backflow prevention assemblies.

As part of this program, a survey of your facility's internal water distribution system was completed on (Month Day, Year). Inspectors reviewed your water distribution system for any piping or connections that could possibly contaminate the water distribution system.

Your facility was either found compliant and/or the necessary changes made to comply with Ordinance. This survey is valid until your facility's next scheduled survey date. You will receive future notice for your next survey date.

If your facility has backflow prevention assemblies requiring testing, you will be receiving additional notice detailing test requirements.

If you have any questions or require additional information, please contact:

_____

Example #3: Cross-Connection Control Program Containment Compliance Notification

A Cross-Connection Control survey was performed at your facility. At that time, it was determined that your facility's potable water system is "contained" by an approved, properly installed backflow prevention device or assembly at the main inlet which is intended to minimize the potential backflow threat to the <Any City USA's> public water system. Therefore, your facility has met the intent of the survey portion of the Cross-Connection Program as defined in Ordinance. Compliance with the survey portion of the program requirements shall remain in effect until your facility's next scheduled survey date.

However, to fully meet the intent of the CCC Program, two (2) items must be addressed:
1. Survey of the facility? **Completed**
2. Successful annual testing of any existing testable backflow prevention assemblies within your facility.

This facility will be in **Compliance** with the Cross-Connection Control Program when the existing backflow prevention assemblies are tested this year and at yearly intervals hereafter. When it is necessary to test such assemblies your facility will receive a notification letter, test forms to be completed by a certified tester for each identified assembly, and a list of certified testers within your facility's area. Upon the successful testing of the backflow prevention assembly, please submit a copy of the completed test record(s) to the water supplier.

Note, however, it is it still possible for existing cross-connections within your facility to potentially affect the water quality within your internal plumbing system. The installation of an approved backflow preventer at the main inlet does not relieve your facility of the responsibility of providing potable water to your employees and the public. In order to comply with all applicable codes and laws, it is recommended that your facility

- Have a cross-connection control survey of the potable water piping system performed within your facility
- Ensure all piping systems downstream of the containment device/assembly are labeled properly
- Ensure backflow prevention assemblies connected to the potable water supply within your facility are tested annually

If you have any questions or require additional information, please contact:

_____

Example #4: Request for Internal Cross-Connection Control Information Notice

The purpose of the <Any City USA's> Cross-Connection Control Program, as defined in Ordinance, is to help eliminate possible contamination of the public water distribution system. There are two required components of the program (1) site survey, and (2) testing of backflow prevention assemblies.

As specified by Ordinance, your facility is required to supply potable water free of existing and/or potential cross-connections to its employees and/or the public. Due to the complexity of your internal piping, a survey of the potable water piping system is necessary to determine if there are any existing and/or potential cross-connections. This survey must be completed by an individual or firm acceptable to the <Any City USA>.

The Potable Water Cross-Connection Survey Report is to be submitted within 30 days from the date of this notice. Accompanied with the Potable Water Piping Cross-Connection Survey Report shall be an Action Plan and timetable for correcting any deficiencies noted in the report.

If you have any questions or require additional information, please contact your Water Supplier at (Phone Number). Your facility's cooperation in this matter is greatly appreciated.

Example #5: Survey Noncompliance Notice 1

The purpose of the <Any City USA's> Cross-Connection Control Program, as defined in Ordinance, is to help eliminate possible contamination of the public water distribution system. There are two required components of the program: (1) site survey, and (2) testing of backflow prevention assemblies.

A survey of your facility's internal water distribution system was completed on (List Date). Inspectors reviewing your water system found connections that could possibly contaminate the public water distribution system. A list of requirements is enclosed.

Requirements on this list must be addressed using only State approved backflow prevention devices. A licensed plumber should be able to assist you with acquiring approved backflow prevention devices. Some backflow prevention devices (assemblies) also require testing by a State Certified Tester. We suggest that the licensed plumber installing the testable assemblies also have the state certification to test assemblies. *All assemblies must be tested immediately at the time of installation.*

These requirements must be completed by (Insert Date). After the requirements and devices have been installed (if applicable), please call the number below on or before the date listed above to schedule a compliance survey. Failure to do so will result in future noncompliance notices.

To arrange for a compliance review or if you require additional information, please contact:

Example #6: Survey Noncompliance Notice 2

The purpose of the <Any City USA's> Cross-Connection Control Program, as defined in Ordinance, is to help eliminate possible contamination of the public water distribution system. There are two required components of the program (1) site survey, and (2) testing of backflow prevention assemblies.

As part of this program, a survey of your facility's internal water distribution system was completed on (insert date) Inspectors reviewing your water system found connections that could possibly contaminate the public water distribution system. A letter of notification was previously sent to you outlining the required corrective measures. For your reference, a duplicate list of requirements is enclosed.

Requirements on this list must be addressed using only State approved backflow prevention devices. A licensed plumber should be able to assist you with acquiring approved backflow prevention devices. Some backflow prevention devices (assemblies) also require testing by a State Certified Tester. We suggest that the licensed plumber installing the testable assemblies also have the state certification to test assemblies. *All assemblies must be tested immediately at the time of installation.*

These requirements must be completed by (insert date). After the requirements and devices have been installed (if applicable), please call the number below on or before the date listed above to schedule a compliance survey. Failure to do so will result in future noncompliance notices.

To arrange for compliance review or if you require additional information, please contact:

_____

Example #7: Cross Connection Control Program Survey Shut-Off Notice

The purpose of the <Any City USA's> Cross-Connection Control Program, as defined in Ordinance, is to help eliminate possible contamination of the public water distribution system.

As part of this program, a survey of your facility's internal water distribution system was completed on (Date: Month Day, Year). Inspectors reviewing your water system found connections that could possibly contaminate the public water distribution system. Two (2) previous letters of notification were sent to you outlining the required corrective measures. For your reference, a duplicate list of requirements is attached.

We presently have no record or notification from you that corrective action has been completed. If you have already completed the requirements, please call the number below to schedule a compliance survey.

You are hereby notified that in accordance with Ordinance, the water supply to the above noted premises will be discontinued as of (Date). Water service may not be resumed until corrective measures have been addressed.

Upon completion of the required corrective action, please contact (**Insert Contact**) on or before the above date at (phone number) to schedule a compliance review.

_____

Example #8: Annual Test Notice

The purpose of the <Any City USA's> Cross-Connection Control Program, as defined in Ordinance, is to help eliminate possible contamination of the public water distribution system. There are two required components of the program: 1) site survey, and 2) testing of backflow prevention assemblies.

This correspondence addresses testing of backflow prevention assemblies, and is independent of previous correspondence pertaining to site survey(s). Periodic testing of backflow prevention assemblies is required to ensure proper working order.

Our records indicate it is time for testing of backflow prevention assemblies at your facility. The enclosed preprinted test forms are the only test forms that will be accepted. Testing should be completed in advance of the completion date noted to allow for repair(s), should they be necessary. Testing of backflow prevention assemblies must be completed by a State approved certified tester. A partial listing is attached for reference.

Following completion of assembly testing and/or repairs, completed test forms may either be faxed to (insert fax number), mailed or emailed to the following address:

<Any City USA>
<Address>

Backflow prevention assemblies within the <Any City USA> are required to be tested on an annual basis. Our records indicate that we have not received the annual test reports on the following backflow assemblies enclosed with this letter.

Completed test forms are to be returned by "[Insert notice response date]". Please retain a copy of the device test results for your records.

If you have any questions or require additional information, please contact:

___

Example #9: Test Notice #2

The purpose of the <Any City USA's> Cross-Connection Control Program, as defined in Ordinance, is to help eliminate possible contamination of the public water distribution system. There are two required components of the program: (1) site survey, and (2) testing of backflow prevention assemblies.

This is your **second notice** pertaining to testing of backflow prevention assemblies, and is independent of previous correspondence pertaining to site survey(s). Periodic testing of backflow prevention assemblies is required to ensure proper working order.

Our records indicate (1) it is time for testing of backflow prevention assemblies at your facility, and that (2) you have not yet returned the previously provided test forms. For your convenience, we have enclosed additional preprinted test forms. Testing of backflow prevention assemblies must be completed by a state registered tester. A partial listing is attached for reference.

Following completion of assembly testing and/or repairs, completed test forms may either be faxed to "[Insert fax number]", or mailed to the following address:

<Any City USA>
<Address>

Completed test forms are to be returned by (insert date). Please retain a copy of the device test results for your records.

If you have any questions or require additional information, please contact:

Example #10: Testing Shut-Off Notice

The purpose of the <Any City USA's> Cross-Connection Control Program, as defined in Ordinance, is to help eliminate possible contamination of the public water distribution system. There are two required components of the program: (1) site survey, and (2) testing of backflow prevention assemblies.

This is your **third notice** pertaining to testing of backflow prevention assemblies, and is independent of previous correspondence pertaining to site survey(s). Periodic testing of backflow prevention assemblies is required to ensure proper working order.

Our records indicate that you have not yet returned the previously provided test forms. For your convenience, we have enclosed additional preprinted test forms. Testing of backflow prevention assemblies must be completed by a State approved certified tester. A partial listing is attached for reference.

You are hereby notified that in accordance with Ordinance, the water supply to the above noted premises will be discontinued as of "[Insert notice response date]". Water service may not be resumed until testing of backflow prevention assemblies has been completed.

Following completion of assembly testing and/or repairs, completed test forms may either be faxed to (insert number), or mailed to the following address:

<Any City USA>
<Address>

Please retain a copy of the device test results for your records. If you have any questions or require additional information, please contact:

_____

Appendix B

Testing Procedures or Methods

INFORMATION PROVIDED BY:

New England Water Works Association, a Section of AWWA

AWWA Pacific Northwest Section

American Society of Sanitary Engineers (ASSE)

University of Florida Training, Research, and Education for Environmental Occupations (TREEO)

The following testing procedures and methods are provided as information to the user. AWWA publication of these testing procedures and methods does not consistute endorsement of any procedure, product or product type, nor does AWWA test, certify, or approve any product. The use of these procedures and methods is entirely voluntary and their use should not supersede or take precedence over or displace any applicable law, regulation, or codes of any governmental authority.

The testing procedures and methods provided may reference material that was not provided to AWWA. The user is directed to contact the appropriate agency for this additional information.

- New England Water Works (NEWWA): http://www.newwa.org/
- AWWA Pacific Northwest Section: http://www.pnws-awwa.org/
- American Society of Sanitary Engineers: http://www.asse-plumbing.org/
- University of Florida Training: http://www.treeo.ufl.edu/

INFORMATION PROVIDED BY:

New England Water Works Association, a Section of AWWA
Three-Valve Differential Test Kit
Field-Test Procedure
Double Check Valve Assembly

This field-test procedure evaluates the operational performance characteristics as specified by nationally recognized industry standards of the independently-operating internal spring loaded check valves while the assembly is in a no-flow condition. This field-test procedure utilizes a three-valve differential pressure test kit to measure the static differential pressure across the check valves. This field-test procedure will reliably detect weak or broken check valve springs and validate the test results by determining that a no-flow condition exists while not closing the upstream shut-off valve. This test procedure will work with all three-valve differential pressure test kits.

Prior to initiating the test, the following preliminary testing procedures shall be followed.
1. The device has been identified.
2. The direction of flow has been determined.
3. The test cocks have been numbered and adapters have been installed.
4. The test cocks have been flushed.
5. Permission to shut-down the water supply has been obtained.
6. The downstream shut-off valve has been closed. (See Note A)
7. The device is inspected and evaluated for a backpressure condition.

The double check valve assembly field-test procedure will be performed in the following sequence to evaluate that:
1. The first check valve has a minimum differential pressure across it of 1 psid.
2. The second valve has a minimum differential pressure across it of 1 psid.
3. The downstream shut-off valve is tight and/or there is no-flow condition through the assembly (including backflow) or no demand downstream.

Note A: Prior to closing the downstream shut-off valve, if it is determined that the device may be prone to backpressure, a standard psi calibrated pressure gauge should be connected to test cock #1 and test cock #4. The pressure readings (psi) should be noted. See Figure B-1.

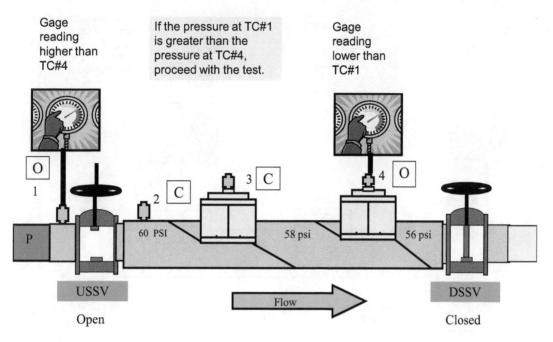

Figure 8-1 Double Check Valve Assembly Test

A. If the pressure (psi) reading at test cock #1 is higher than the pressure (psi) reading at test cock #4, close the downstream shut-off valve and proceed to Step 1, number 3.
B. If the pressure (psi) reading at test cock #1 is lower than the reading at test cock #4, the device is in a backpressure condition and the downstream shut-off valve must be closed prior to performing the test of the device. See Figure B-2.
 1. After closing the downstream-shutoff valve, test cock #4 should be bled again and the pressure readings at test cock #1 and #4 should be noted. If the pressure reading at test cock #1 is higher than the reading at test cock #4, proceed to Step 1, number 3. If the pressure reading at test cock #1 is still lower than the reading at test cock #4, the downstream shut-off valve is considered leaking and a backpressure condition still exists. The downstream shut-off valve must be reclosed, repaired, or a no-flow condition must be established before testing the device. The device cannot be tested in a backpressure condition.

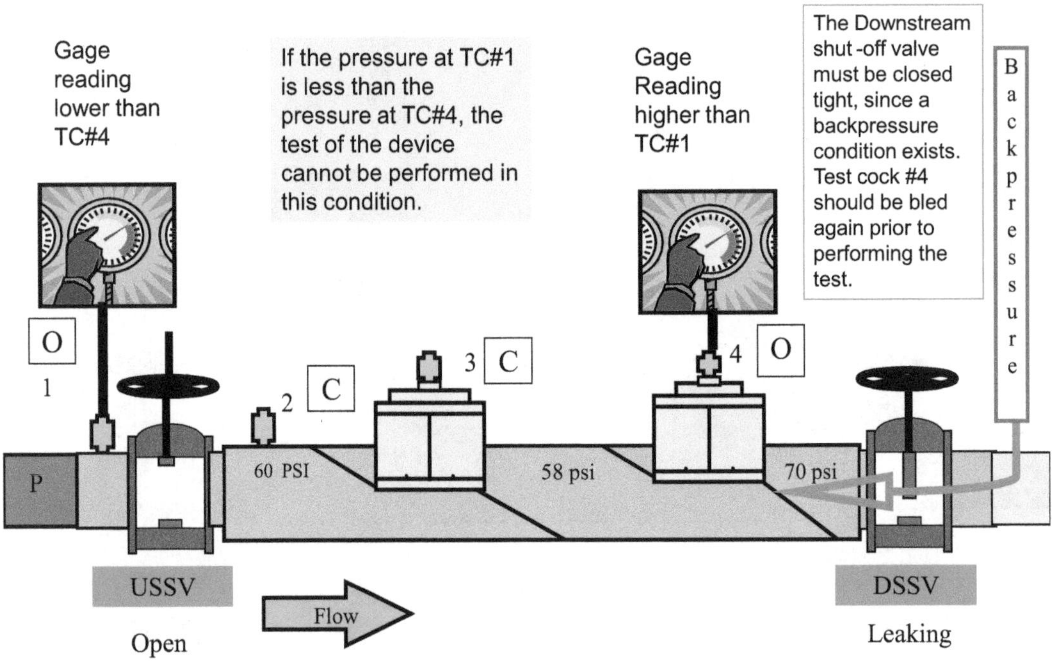

102 BACKFLOW PREVENTION AND CROSS-CONNECTION CONTROL

Figure 8-2 Double Check Valve Assembly Test—Backpressure Condition

DOUBLE CHECK VALVE ASSEMBLY THREE VALVE FIELD-TEST PROCEDURE

Step 1: Test the first check valve to determine that it has a minimum static differential pressure across it of 1 psi (See Figure B-3.)

1. Verify that upstream shut-off valve is open.
2. Close the downstream shut-off valve (If it is determined that the device is prone to backpressure as in a fire protection system, see NOTE A prior to closing the downstream shut-off valve.)
3. Orientate the test kit. Close high and low control valves on the test kit. Open the vent control valve.
4. Connect the high pressure hose to test cock #2.
5. Connect the low pressure hose to test cock #3.
6. Open test cocks #2 and #3.
7. Open the high control valve on the test kit to bleed the air from the high pressure hose. Close the high control valve. (Water will bleed through the vent hose.)
8. Open the low control valve on the test kit to bleed the air from the low pressure hose. Close the low control valve. (Water will bleed through the vent hose.)
9. The differential pressure gauge reading should be a minimum of 1 psid. This differential pressure gauge reading is the apparent reading. This gauge reading cannot be validated until it is confirmed that the device is in a no-flow condition. (See NOTE B)
10. Close test cocks #2 and #3. Disconnect the hoses.

NOTE B: If the differential pressure is 0 psid, this is an indication that the first check valve is leaking and the device and downstream-off valve cannot be tested for tightness using the procedure outlined in Step 3. However, an affirmation can be made that since the first check valve has a differential pressure of 0 psid, the device is in a no-flow condition. The gauge would record a positive psid if the device was in a flow condition. The second check valve can and should be tested to determine if the device is providing protection.

AWWA Manual M14

TESTING PROCEDURES OR METHODS   103

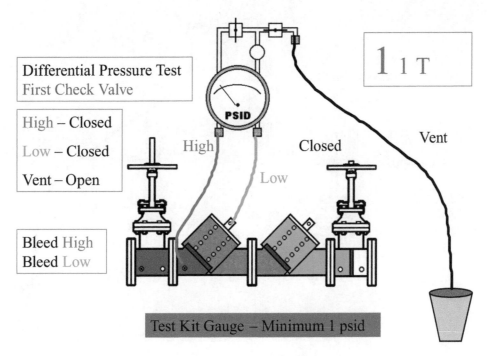

Figure 8-3   DCVA—Step #1

***Step 2: Test second check valve to determine that it has a minimum static differential pressure differential across it of 1 psi** (See Figure B-4.)*

1. Orientate the test kit valves. Close high and low control valves. Open vent control valve.
2. Connect the high pressure hose to test cock #3.
3. Connect the low pressure hose to test cock #4.
4. Open test cocks #3 and #4.
5. Open the high control valve on the test kit to bleed the air from the high pressure hose. Close the high control valve.
6. Open the low control valve on the test kit to bleed the air from the low pressure hose. Close the low control valve.
7. The differential pressure gauge reading should be a minimum of 1 psid. The differential pressure gauge reading is the apparent reading. This gauge reading cannot be validated until it is confirmed that the device is in a no-flow condition. (See Note C)
8. Close tests cocks #3 and #4. Disconnect the hoses.

Note C: If the differential pressure is 0 psid, this is an indication that the second check valve is leaking if the device is confirmed to be in a no-flow state (no backpressure). The device and downstream shut-off valve cannot be tested for tightness using the procedure outlined in Step 3. However, the device should be tested for backpressure, since a 0 psid reading across the second check valve may be an indication that the downstream shut-off valve is leaking and the device is in a backflow condition.

AWWA Manual M14

104 BACKFLOW PREVENTION AND CROSS-CONNECTION CONTROL

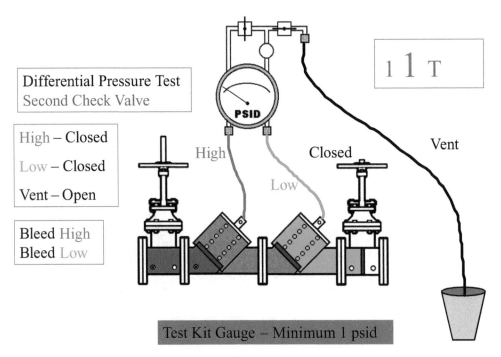

Figure 8-4 DCVA—Step #2

Step 3: Tightness validation test—test the device for no-flow. (See Figure B-5.)

To test the device for no-flow, both check valves must be tight and holding a minimum differential pressure of 1 psid. There must be little or no fluctuation of inlet supply pressure. Any backpressure situation should be evaluated. The upstream shut-off valve is open and the downstream shut-off valve is closed.

1. Orientate test kit valves. Close high and low control valves. Open vent control valve.
2. Connect the high pressure hose to test cock #2 and the low pressure hose to test cock #3.
3. Open test cocks #2 and #3.
4. Open the high control valve on the test kit to bleed air from the high pressure hose. (Water will discharge out of the vent host.) Close the high control valve.
5. Open the low control valve on the test kit to bleed air from the low pressure hose. (Water will discharge out of the vent host.) Close the low control valve.
6. The differential pressure gauge reading should be a minimum of 1 psid.
7. Elevate the vent hose and open the low control valve to fill vent hose with water. Close the low control valve and connect the vent hose to test cock #4. Open test cock #4.
8. Open the test kit high control valve. (This supplies high pressure water downstream of check valve number 2.) If the differential pressure rises, close test cock #4 immediately. (See NOTE D)
9. Close test cock #2. (This stops the supply of high pressure water to the test kit gauge and downstream of check valve number 2.)
10. Observe the test kit needle. If the differential pressure gauge reading holds steady, the device is recorded as being under a no-flow condition. (See NOTE E) If the differential pressure gauge reading drops to zero, the device is in a flow condition and downstream shut-off valve is recorded as leaking. (See NOTE F)

TESTING PROCEDURES OR METHODS 105

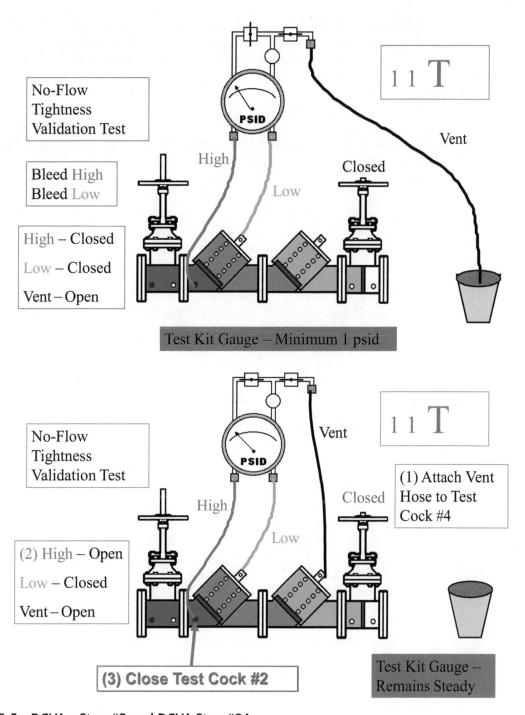

Figure 8-5   DCVA—Step #3 and DCVA Step #3A

NOTE D: If a backpressure condition is present with a leaking downstream shut-off valve and with the high and vent control valves open, nonpotable water will pass through the test kit and be introduced into the potable water supply. If this occurs, test cock #4 should be closed immediately, the test should be discontinued, and the test kit should be removed and flushed-out with potable water. The assembly should be tested for backpressure as stated above and retested making sure that the downstream shut-off valve is closed tight or no-flow can be achieved and validated.

NOTE E: To determine the tightness of the downstream shut-off valve, a demand downstream of the backflow prevention device assembly shall be created while performing the no-flow test. If the needle on the test kit remains steady during a demand condition, the downstream shut-off valve is considered holding tight. If under a demand condition the needle on the test kit drops to zero, the downstream shut-off valve is considered leaking. If there is no water demand downstream of the backflow prevention device assembly, the tightness validation of the downstream shut-off valve may not be possible, since a leaking downstream shut-off valve with a no-flow condition will emulate a tight downstream-shutoff valve.

NOTE F: With a leaking downstream shut-off valve, the device is in a flow condition and the previous readings taken are invalid. The device does not fail the test, since it cannot be tested in a flow condition. To proceed with the test of the device, a no-flow condition shall be achieved, either through the repair of the downstream shut-off valve, the operation of an additional shut-off valve downstream, or by another means of validating that the device is under a no-flow condition.

Concluding Procedures: This completes the standard field test for a double check valve assembly. Before removing the test equipment, the tester should ensure that all test cocks have been closed and the downstream shut-off valve is open, thereby re-establishing flow. All test data should be recorded on appropriate forms.

New England Water Works Association
Three-Valve Differential Test Kit
Field-Test Procedure
Pressure Vacuum Breaker

This field test procedure evaluates the operational performance characteristics as specified by nationally recognized industry standards of the independently operating internal spring loaded check valve and air inlet valve while the assembly is in a no-flow condition. This field-test procedure utilizes a three-valve differential pressure test kit to measure the static differential pressure across the check valve and determine the opening point of the air inlet valve. This field-test procedure will reliably detect weak or broken check valve springs and validate the test results by determining that a no-flow condition exists. This test procedure will work with all three-valve differential pressure test kits.

Prior to initiating the test, the following preliminary testing procedures shall be followed.
1. The device has been identified.
2. The direction of flow has been determined.
3. The test cocks have been numbered and the hood is removed.
4. Test adapters have been installed and "blown-out."
5. Permission to shut down the water supply has been obtained.
6. The downstream shut-off valve has been closed.

This test procedure will examine the pressure vacuum breaker assembly for the following performance characteristics using a three-valve differential pressure gauge with a range of 0–15 psid.
1. The check valve has a minimum differential pressure across it of 1 psid.
2. The downstream shut-off valve is closed tight and/or a no-flow condition exists.
3. The air inlet valve opens at least 1 psid above atmospheric pressure.

TESTING PROCEDURES OR METHODS

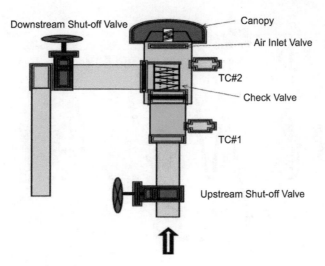

Figure 8-6 Pressure vacuum breaker

PRESSURE VACUUM BREAKER THREE-VALVE FIELD-TEST PROCEDURE

Step 1: Test the check valve to determine that it has a minimum differential pressure across it of 1 psid (see Figure B-7.)

1. Verify that the upstream shut-off valve is open.
2. Close the downstream shut-off valve.
3. Orientate test kit. Close high and low control valves. Open vent control valve.
4. Connect the high pressure hose to test cock #1.
5. Connect the low pressure hose to test cock #2.
6. Place the vent hose into a bucket or suitable drainage area.
7. Open test cock #1 and test cock #2.
8. Open the high control valve; bleed water through the vent hose.
9. Close high control valve.
10. Open the low control valve; bleed water through the vent hose.
11. Close low control valve.
12. Observe needle on test kit. Should be 1 psid or greater. The differential pressure gauge reading is the apparent reading. This gauge reading cannot be validated until it is confirmed that the device is under a no-flow condition.
13. Shutoff test cock #1 and #2.
14. Remove hoses from the device.
15. Proceed to Step 2.

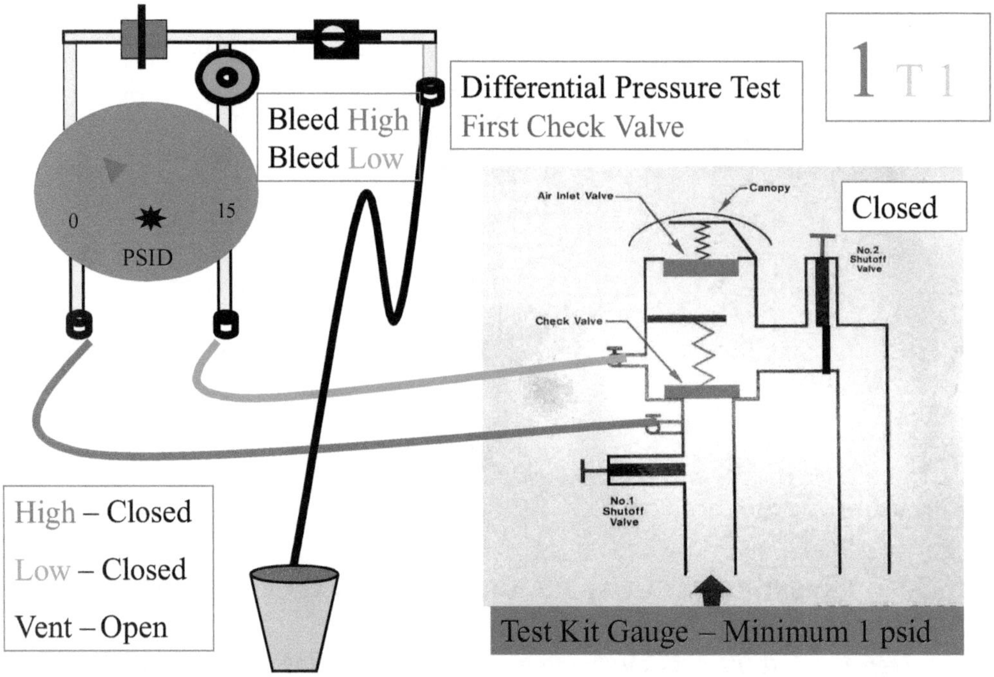

Figure 8-7 PVB—Step #1

Step 2: Tightness validation test—no-flow test to determine that the device is under a no-flow condition and validate differential pressure reading (See Figure B-8.)

1. Downstream shut-off valve remains closed and upstream shut-off valve remains open.
2. Place low pressure and vent hoses in a bucket or suitable drainage area.
3. Connect high pressure hose to test cock #2.
4. Position the test kit valves: high and low control valve closed; vent control valve open.
5. Open test cock #2.
6. The test kit needle should "peg" to the extreme right of the gauge.
7. Open high control valve to bleed air; close the high control valve.
8. Close the upstream shut-off valve. (This stops the supply of high-pressure water to the device and test kit gauge.)
9. Observe needle on test kit. If the needle remains steady, the device is in a no-flow condition. If the needle starts to descend to zero, the device is in a flow condition and the downstream shut-off valve is considered leaking. (See Note A)
10. Proceed to Step 3 if a no-flow condition exists.

NOTE A: If the device is in a flow condition, the differential reading taken is invalid. The device does not fail the test; it cannot be tested since it is in a flow condition. To perform the test of the device, a no-flow condition shall be achieved, either through the repair of the downstream shut-off valve, the operation of an additional shut-off valve downstream, or by another means of validating that the device is under a no-flow condition.

Downstream Shut-off Valve Tightness: To determine the tightness of the downstream shut-off valve, a demand downstream of the backflow prevention device assembly shall be created while performing the no-flow test. If the needle on the test kit remains steady during a demand condition, the downstream shut-off valve is considered holding tight. If under a demand condition the needle on the test kit drops to zero, the downstream shut-off valve is considered leaking. If there is no water demand downstream of the backflow prevention device assembly, the tightness validation of the downstream shut-off valve may not be possible, since a leaking downstream shut-off valve with a no-flow condition will emulate a tight downstream-shutoff valve.

Step 3: Determine if the air inlet valve opens at least 1 psid above atmospheric pressure (See figure B-9.)

1. Both shut-off valves are still closed.
2. The high pressure hose is still connected to the open test cock #2.
3. The low pressure and vent hoses are still in a bucket or suitable drainage area.
4. The test kit valves are positioned as follows: high and low control valve closed; vent control valve is open.
5. Elevate the test kit and the end of low pressure hose to the same level as the air inlet Valve.
6. Slowly open the high control valve while simultaneously observing the air inlet valve. (Lightly placing an object on top of the air inlet may be helpful in determine the opening point.)
7. Observe the test kit needle at the point where the air inlet valve opens (pops). The air inlet should open at a minimum of 1 psid or greater. If the air inlet valve does not open, the upstream shut-off valve may be leaking or air inlet valve is stuck closed.
8. Observe the air inlet valve to determine that it is open completely.

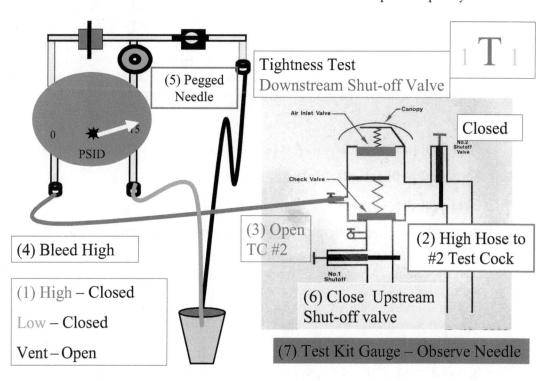

Figure 8-8   PVB—Step #2

NOTE: Numbers on illustration may not correlate with the step numbers above.

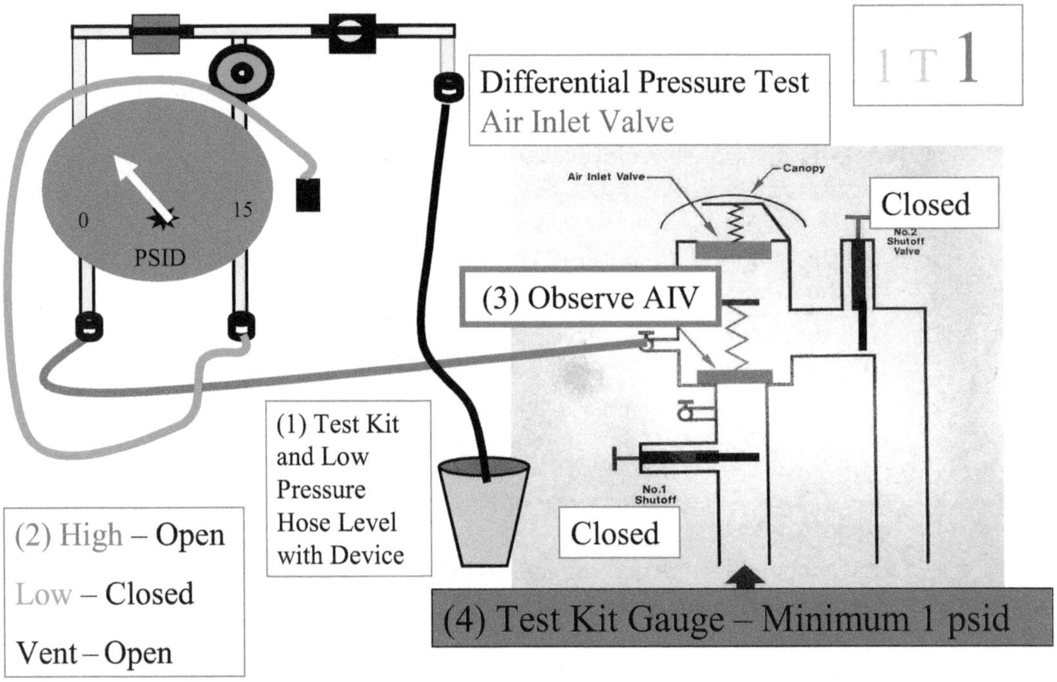

Figure 8-9 PVB—Step #3

NOTE: Numbers on illustration do not correlate with the step numbers above.

Concluding Procedures: This completes the standard field test for a pressure vacuum breaker. Before removal of the test equipment, the tester should ensure that the test cocks have been closed, and the downstream and upstream shut-off valves are open, thereby reestablishing flow. All test data should be recorded on appropriate forms.

New England Water Works Association
Three-Valve Differential Test Kit
Field-Test Procedure
Reduced-Pressure Principle Backflow Prevention Assembly (RPZ)

This field-test procedure evaluates the operational performance characteristics as specified by nationally recognized industry standards of the two independently operating internal spring loaded check valves and a mechanical, independently operating, hydraulically dependent relief valve located between the check valves while the assembly is in a no-flow condition. This field-test procedure utilizes a three valve differential pressure test kit to evaluate the tightness of the both the first and second check valves, measure the static differential pressure across the first and second check valves, and test the operation of the relief valve. This field test procedure will reliably detect weak or broken check valve springs and validate the test results by determining that a no-flow condition exists while not closing the upstream shut-off valve. This test procedure will work with all three-valve differential pressure test kits.

Prior to initiating the test, the following preliminary testing procedures shall be followed.

1. The device has been identified.
2. The direction of flow has been determined.
3. The test cocks have been numbered and adapters have been installed.
4. The test cocks have been flushed. (See NOTE A)
5. Permission to shut down the water supply has been obtained.
6. The downstream shut-off valve has been closed.
7. No water is discharging from the relief valve opening.

The reduced pressure principle backflow prevention assembly field-test procedure will be performed in the following sequence to evaluate that:

1. The first check valve will be tested to determine tightness and a minimum differential pressure across the first valve of 5.0 psid.
2. The second check valve will be tested to determine tightness against backpressure and a minimum differential pressure across the second check valve of 1.0 psid.
3. The downstream shut-off valve will be tested for tightness and/or the device is in a no-flow condition at the time of the test.
4. The relief valve will be tested to determine if the relief valve opens at a minimum differential pressure of 2 psid below the inlet supply pressure.

NOTE A: When flushing the test cocks on a reduced-pressure principle assembly, test cock #4 should be flushed first and left open with a small amount of flow, while flushing test cocks #1, #2, and #3. Once test cocks #1, #2, and #3 have been flushed, close test cock #4. This prevents the premature opening of the relief valve prior to the test.

REDUCED-PRESSURE PRINCIPLE BACKFLOW PREVENTION DEVICE ASSEMBLY—THREE-VALVE FIELD-TEST PROCEDURE

Step 1: Test the first check valve to determine if it is tight and has a minimum differential pressure across it of 5 psid (See Figure B-10.)

1. Verify that the upstream shut-off valve is open.
2. Close the downstream shut-off valve. If no water discharges from the relief valve, the first check valve is considered tight; proceed with the test. If water discharges from the relief valves, the first check valve is considered leaking and it must be repaired prior to completing the test.
3. Orientate the test kit; close the high and low control valves on test kit. Open the test kit bleeder/vent control valve.
4. Connect the high pressure hose to test cock #2.
5. Connect the low pressure hose to test cock #3.
6. Open test cocks #2 and #3.
7. Open the high control valve on the test kit to bleed the air from the high pressure hose.
8. Close the high control valve.
9. Open the low control valve on the test kit to bleed the air from the low pressure hose.
10. Close the low control valve.
11. The differential pressure gauge reading should be a minimum of 5 psid. This differential pressure gauge reading is the apparent reading, and it cannot be validated until it is confirmed that the device is in a no-flow condition.

112 BACKFLOW PREVENTION AND CROSS-CONNECTION CONTROL

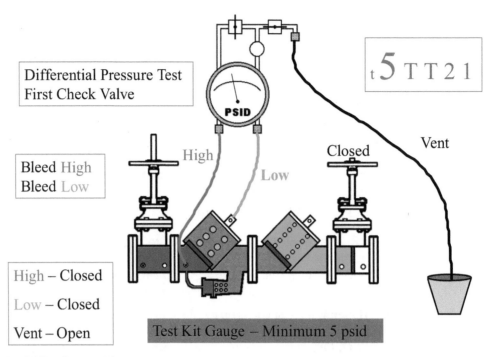

Figure 8-10 RPZ—Step #1A

Step 2: Test the tightness of the second check valve against backpressure (See Figure B-11.)

1. The test kit valves and hoses are in the same position as at the conclusion of Step 1. (High and low control valves are closed and the vent control valve is open.)
2. Elevate the vent hose and open the low control test kit valve to fill vent hose with water. Close the low control test kit valve.
3. Connect the water filled vent hose to test cock #4 and open test cock #4.
4. Open the high control test kit valve. (This supplies high-pressure water to the downstream side of second check valve.) If the differential pressure rises, close test cock #4 immediately. (See Note C) The second check valve is considered tight if the differential pressure gauge remains steady and no water is discharging from the relief valve. If the differential pressure gauge reading on the test kit drops and water discharges from the relief valve, the second check is leaking. (See Note B)

Note B: If the second check valve is leaking, the downstream shut-off valve and/or the no-flow test (Step 3) cannot be performed. However, an affirmation can be made that since water is discharging from the relief valve, the downstream shut-off valve is considered tight or the device is in a no-flow condition. The deferential pressure test across the second check valve (Step 5) should still be performed since it may reveal a failed check or O-ring problem. If the differential pressure is 0 psid and backpressure is not present (See Step 5, #9) it is most likely a check valve failure. If the differential is positive, it is most likely an O-ring problem. The relief valve can and should be tested. To test the relief valve with a failed second check valve, close test cock #4 and proceed to Step 4.

AWWA Manual M14

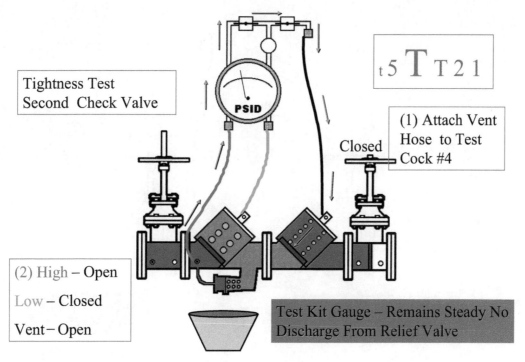

Figure 8-11 RPZ—Step #2

NOTE C: If the differential pressure reading on the test kit increases when the high control test kit vales is opened (as stated in Step 2, Number 4), the device may be in a backflow condition with a leaking downstream shut-off valve. (High pressure water is flowing back through the leaking downstream shut-off valve, through test cock #4, through the test kit to test cock #2 and into the potable water supply.) If this occurs, test cock #4 should be closed immediately and the test should be terminated. The test kit should be removed and flushed out with potable water. Attempts should be made to reclose the downstream shut-off valve and a pressure gauge should be used to test for backpressure prior to retesting the assembly. The downstream shut-off valve should be confirmed closed or a no-flow condition achieved and validated prior to conducting the test of the assembly. (The downstream shut-off valve may need to be repaired or replaced.)

Step 3:   Tightness validation test—test the device for no-flow to determine that the device is under a no-flow condition and validate differential pressure readings (See Figure B-12.)

1. The test kit valves and hoses are positioned as at the conclusion of Step 2.
2. **Close test cock #2.** (This stops the supply of high-pressure water to the test kit and downstream of the second check valve.)
3. Observe the test kit needle. If the differential pressure gauge reading holds steady, the downstream shut-off valve is tight and/or the device is under a no-flow condition. If the differential pressure gauge drops to zero, the downstream shut-off valve is leaking and the device is in a flow condition. (See NOTE D)
4. **Open test cock #2.**

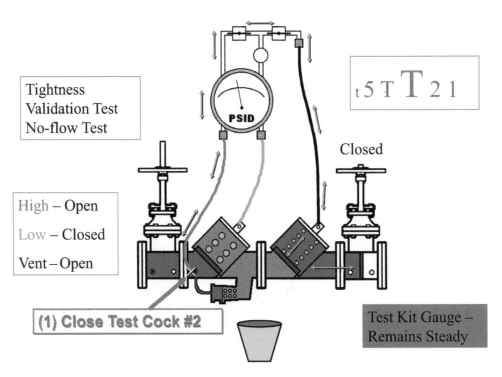

Figure 8-12   RPZ—Step #3

NOTE D: If the device is in a flow condition, the psid readings previously taken are invalid, and the device must be retested once a no-flow condition can be achieved. The device does not fail the test, since it cannot be tested in a flow condition. A no-flow condition shall be achieved, either through the repair of the downstream shut-off valve, the operation of an additional shut-off valve downstream, or by another means of validating that the device is in a no-flow condition. A compensating temporary bypass hose may be used in some cases. (See NEWWA Use of a bypass Hose in Reduced-Pressure Principle Backflow Prevention Device Testing).

Downstream Shut-off Valve Tightness: To determine the tightness of the downstream shut-off valve, a demand downstream of the backflow prevention device assembly shall be created while performing the no-flow test. If the needle on the test kit remains steady during a demand condition, the downstream shut-off valve is considered holding tight. If under a demand condition the needle on the test kit drops to zero, the downstream shut-off valve is considered leaking. If there is no water demand downstream of the backflow prevention device assembly, the tightness validation of the downstream shut-off valve may not be possible, since a leaking downstream shut-off valve with a no-flow condition will emulate a tight downstream-shutoff valve.

Step 4:   Test the relief valve to determine that it opens at a minimum differential pressure of 2 psid below the inlet supply pressure (See Figure B-13.)

1. The test kit valves and the hoses are positioned as at the conclusion of Step 3. Test cock #2 should be open.
2. Slowly open the test kit low control needle valve ¼ **turn**.
3. Record the differential pressure gauge reading at the point when water initially drips from the relief valve opening. The differential pressure gauge reading should be a minimum of 2 psid. If water does not discharge from the relief valve, it may be jammed (intentionally), the sensing line may be clogged, or the diaphragm cannot open due to mechanical wear.

TESTING PROCEDURES OR METHODS   115

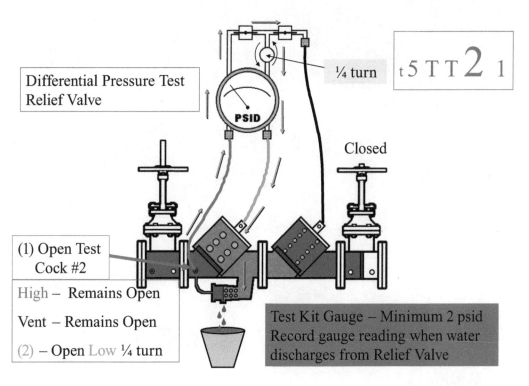

Figure 8-13   RPZ—Step #4

Step 5:   Test the second check valve differential pressure **(See Figure B-14)**

Testing the differential pressure across the second check valve will validate the tightness of the downstream shut-off valve and determine if a backpressure condition exists. If the downstream valve is leaking and the device is in a flow condition, the differential pressure test across the second check valve cannot be performed.

1. Orientate the test kit; close high and low control valves. Open the vent control valve.
2. Connect the high pressure hose to test cock #3.
3. Connect the low pressure hose to test cock #4.
4. Open test cocks #3 and #4.
5. Open the high control valve on the test kit to bleed the air from the high pressure hose.
6. Close the high control valve.
7. Open the low control valve on the test kit to bleed the air from the low pressure hose.
8. Close the low control valve.
9. Record the differential pressure gauge reading. It should be a minimum of 1 psid, if the second check valve was held tight against backpressure. If the differential pressure reading across the second check valve is 0 psid, this is an indication that the second check valve spring is damaged or the downstream shut-off valve is leaking and the device is under a backpressure condition. Evaluate for backpressure to eliminate this possibility. (See evaluation of backpressure when testing a DCVA)

116 BACKFLOW PREVENTION AND CROSS-CONNECTION CONTROL

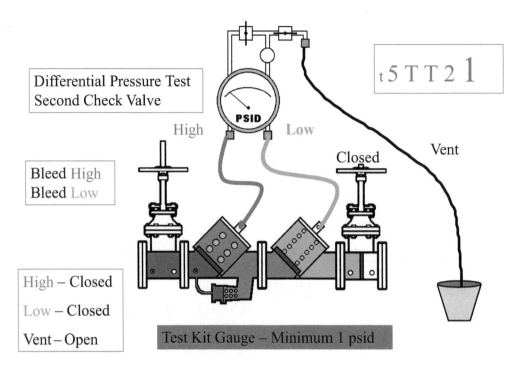

Figure 8-14 RPZ—Step #5

Concluding Procedures: This completes the standard field test for a reduced-pressure principle backflow prevention device. Before removal of the test equipment, the tester should ensure that the test cocks have been closed, and the downstream shut-off valve is open, thereby re-establishing flow. All test data should be recorded on appropriate forms.

New England Water Works Association
Three-Valve Test Kit
Field-Test Procedure
Spill-Resistant Pressure Vacuum Breaker

This field-test procedure evaluates the operational performance characteristics as specified by nationally recognized industry standards of the independently operating internal spring loaded check valve and air inlet valve while the assembly is in a no-flow condition. This field-test procedure utilizes a three-valve differential pressure test kit to measure the static differential pressure across the check valve and determine the opening point of the air inlet valve. This field-test procedure will reliably detect weak or broken check valve springs and validate the test results by determining that a no-flow condition exists. This test procedure will work with all three-valve differential pressure test kits. (See Fig. B-15.)

Prior to initiating the test, the following preliminary testing procedures shall be followed:
1. The device has been identified.
2. The direction of flow has been determined.
3. The test cocks have been numbered and the canopy is removed.
4. A test adapter has been installed and "blown-out."
5. Permission to shut down the water supply has been obtained.

AWWA Manual M14

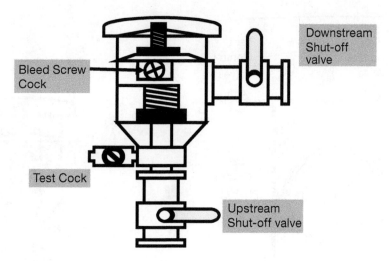

Figure 8-15 Spill-resistant pressure vacuum breaker

6. The downstream shut-off valve has been shutoff.
   This test procedure will examine the spill-resistant pressure vacuum breaker assembly for the following performance characteristics using a three-valve differential pressure gauge with a range of 0 – 15 psid.
1. The check valve has a minimum differential pressure across it of 1 psid.
2. The downstream shut-off valve is closed tight and/or a no-flow condition exists.
3. The air inlet valve opens at least 1 psid above atmospheric pressure.

SPILL-RESISTANT PRESSURE VACUUM BREAKER THREE-VALVE FIELD-TEST PROCEDURES

Step 1: Test the check valve to determine that it has a minimum differential pressure across it of 1 psid (See Figure B-16.)

1. Verify that the upstream shut-off valve is open.
2. Close the downstream shut-off valve.
3. Orientate the test kit valves—high and low control valves closed; vent control valve open.
4. Connect the high hose to the test cock.
5. Open the test cock. The test kit needle should peg to the extreme right of the gauge.
6. Open high control valve to bleed air from the hose; close the high control valve.
7. Close the upstream shut-off valve.
8. Raise test kit and end of the low pressure hose to the elevation of the air inlet valve.
9. *Slowly* unscrew the bleed screw until it starts to drip.
10. When dripping from the bleed screw stops, and the needle on the test kit stabilizes, record the differential pressure. *It must be 1 psid or greater.* If water continues to flow from the bleed screw, the upstream shut-off valve may be leaking. The differential pressure gauge reading is the apparent reading. This gauge reading cannot be validated until it is confirmed that the device is under a no-flow condition.
11. Close the bleed screw.

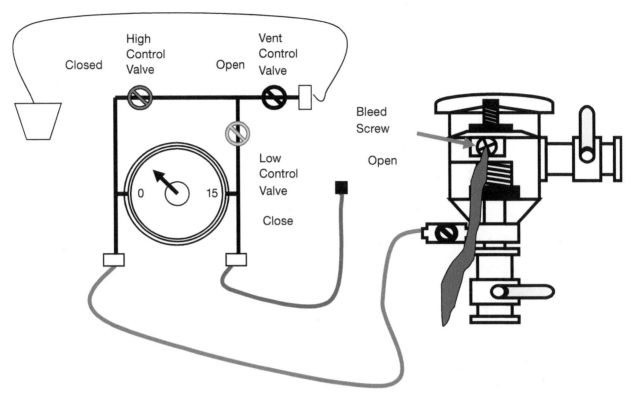

Figure 8-16   Step 1

Step 2: Tightness validation test—test the device to determine if the device is under a no-flow condition and validate differential pressure reading (See Figure B-17.)

1. With the high pressure hose still connected to the test cock, open the upstream shut-off valve to pressurize the device. The test kit needle should peg to the extreme right of the gauge. The downstream shut-off valve is still closed.
2. Open the high control valve to bleed air from the hose; close the high control valve.
3. Close the upstream shut-off.
4. Observe needle on the test kit. If the needle remains steady the downstream shut-off valve is holding tight and/or the device is under a no-flow condition. If needle starts to descend, the downstream shut-off valve is considered leaking (See Note A).
5. Record data.
6. Proceed to Step 3 if a no-flow condition exists.

Note A: If the device is in a flow condition the differential reading taken is invalid. The device does not fail the test, since it cannot be tested in a flow condition. To perform the test of the device, a no-flow condition shall be achieved, either through the repair of the downstream shut-off valve, the operation of an additional shut-off valve downstream, or by another means of validating that the device is under a no-flow condition. To determine the condition of the downstream shut-off valve, a demand downstream must be created during the no-flow test. If during a created demand the needle on the test kit continues to hold steady, the downstream shut-off valve is considered tight.

TESTING PROCEDURESOR METHODS 119

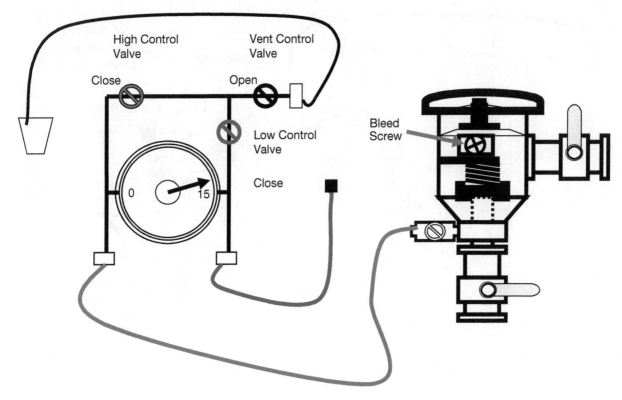

Figure 8-17   Step 2

Step 3: Determine if the air inlet valve opens at least 1 psid above atmospheric pressure (See Figure B-18.)

1. Both shut-off valves are still closed and the canopy is removed.
2. The high pressure hose is still connected to the open test cock.
3. The test kit valves are positioned as follows: high and low control valves are closed; vent control valves open.
4. Hold the test kit and end of the low pressure hose at the same level as the air inlet valve.
5. Slowly unscrew the bleed screw until it starts to drip.
6. Slowly open the high control valve ¼ turn while simultaneously observing the air inlet valve. (Lightly placing an object on top of the air inlet may be helpful in determine the opening point.)
7. Read the test kit needle at the point where the air inlet valve opens (pops). It should be equal to or greater than 1 psid. A reading of less than 1 psid is cause for failure. If the air inlet valve does not open, the upstream shut-off valve may be leaking.
8. Observe the air inlet valve to determine that it is open completely.

Concluding procedures: This completes the standard field test for a spill-resistant pressure vacuum breaker. Before removal of the test equipment, the tester should ensure that the bleed screw and test cock are closed, and the downstream and upstream shut-off valves are open, thereby re-establishing flow. All test data should be recorded on appropriate forms and submitted to the appropriate parties.

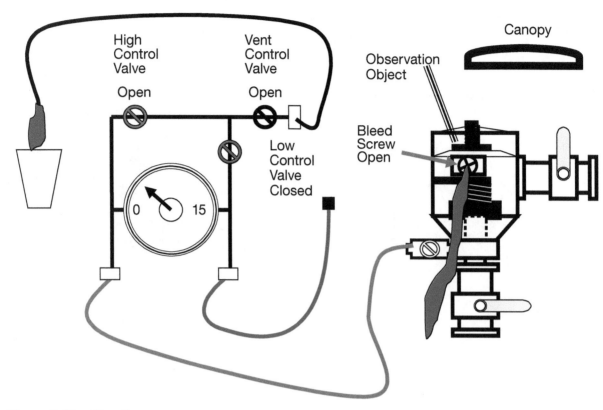

Figure 8-18   Step 3

INFORMATION PROVIDED BY:

AWWA Pacific Northwest Section

Backflow Assembly Testing

This section discusses the following topics:

- Tester responsibilities
- Tester ethics
- Safety considerations for backflow assembly testers and specialists.
- Description of general backflow test procedures
- Description of backflow test equipment

Introduction

By becoming familiar with these aspects of testing and the expectations for each, cross-connection specialists will be better equipped to develop quality assurance or testing programs.

Backflow assembly *testing is an important element* of a cross-connection control program. Competent testers help to ensure that properly functioning assemblies are protecting the water distribution system from contamination or pollution caused by backflow.

Backflow assembly testers have important responsibilities to their customers and to the water supplier. Testers should perform their work by using

- Approved test procedures.
- Test equipment that is properly maintained and verified for accuracy.

- Appropriate safety equipment and following the safety requirements of the administrative authority.

Testers may also need to meet
- Backflow assembly tester training, certification, or licensing requirements.
- Other federal, state, or local rules such as business licensing, insurance, and additional registrations or licensing.
- Specific requirements of the water supplier, such as in the quality assurance and quality control program for backflow assembly testing.

> *Testing is an important element* of a cross-connection control program to protect the water distribution system.

Backflow Assembly Tester Training and Certification

Many states or other jurisdictions require backflow assembly testers to be certified or licensed to demonstrate competency. These states or other jurisdictions may issue certifications or licenses. They may also accept certifications or licenses from other states, jurisdictions, or recognized industry organizations, such as AWWA, ABPA, ASSE, or ABC. Training a person to competently test backflow assemblies is generally from an approved facility or instructor.

Many states or other jurisdictions require testers to periodically renew their certifications or licenses. This may be accomplished by completing continuing education courses and by demonstrating competency through written and hands-on exams. Continuing education and certification or license renewal requirements are recommended for backflow assembly testers every two to three years.

Update training and practical hands-on exams serve to reinforce the use of proper test procedures. This keeps testers up-to-date with current industry information and regulatory requirements.

> Prior to testing, backflow assembly testers should check for *requirements in the jurisdiction in which they are working.*
>
> *Test kits* are also referred to as test equipment and include a gauge, hoses, valves, and other fittings.

Backflow Assembly Tester Responsibilities

The backflow assembly tester has a responsibility to:
- Meet all certification, licensing and other *requirements in the jurisdiction* where they are providing testing services.
- Be properly equipped and capable of safely and properly using tools, *test kits,* and other equipment needed to test assemblies and inspect air gaps.
- Maintain tools, test kits, and other equipment according to manufacturer or regulatory requirements.
- Properly test backflow assemblies using test procedures accepted/pproved by the administrative authority.
- Accurately and completely report the test results.

> *Some jurisdictions may require approval of test equipment.*

> The BAT should inspect all backflow assemblies for proper:
> - Orientation
> - Approval status
> - Installation

> Some testers may not want to submit test reports until they receive payment for services.

> Verify any test report form requirements with the *water supplier or administrative authority.*

> Backflow assembly testers should *report conditions* on the test report form such as:
> - *Failed tests*
> - *Repairs performed*
> - *Incorrect installations or inadequate clearances for access, testing or repairs*
> - *Unapproved assembly or orientation*
> - *Incorrect type of assembly for the hazard, such as a DC when the supplier requires an RP*
> - *Missing or damaged components such as test cocks, plugs, caps or shutoff valves*
> - *Flooded assemblies or inadequate drainage*
> - *Any hazards or adverse conditions that could affect the ability of an assembly to prevent backflow*

- Provide test reports to the water supplier or other administrative authority, if required, and to the owner of the assembly. Retain copies for as long as required by the administrative authority.

Some testers may need to meet additional business-related requirements in order to perform their work on customer's premises. Some jurisdictions may require a tester to obtain a business or contractor license, as well as carry various types of insurance. This may include proof of business, vehicle, liability, bond, or other such types of insurance.

Accurate backflow assembly test results are obtained using approved test procedures and calibrated test equipment. The approval of test equipment by type or manufacturer may be required in some states or jurisdictions. Some states or jurisdictions require testers to periodically (often annually) verify test equipment for accuracy. Whenever test equipment damage is suspected or inaccurate readings are obtained, test equipment recalibration, repair, or replacement may be necessary. See the section in this chapter on test equipment for additional information.

Responsibilities to the Water Supplier

The backflow assembly tester has a responsibility to the water supplier to

- Properly test backflow assemblies that protect the public water system according to approved procedures and other requirements.
- Submit complete and accurate test reports to the water supplier within a reasonable amount of time (such as within 10 days of test completion) or as specified by the supplier.
- Provide documentation, such as tester certification or licensing documents, test equipment accuracy reports or calibration reports, and other information as requested to meet the quality assurance element of the supplier's cross-connection control program.

See the *test report forms* section in this chapter for additional information.

Responsibilities to the Customer

The backflow assembly tester has a responsibility to the customer to

- Properly test customer-owned backflow assemblies using approved test procedures and calibrated equipment.
- Submit complete and accurate test reports to the customer and to the water supplier on the customer's behalf.
- Notify occupants in advance if water service must be shutoff for the test.
- Restore water service to the same status as originally found and note any possible discrepancies on test report form.

Some jurisdictions or customers may have additional notification requirements before testing or shutting off water. For example, an assembly on a fire system may have specific shutoff procedures that must be followed so that alarms are not sent to monitoring or emergency services or the fire suppression system is not activated. Manufacturing plants may need advance notice to plan testing during nonproduction hours.

The tester should follow common customer service practices. These include answering questions about backflow and testing procedures, being careful of the customer's property, and providing the customer a copy of the test report form.

Backflow Assembly Test Report Forms

Test reports are legal public health records. They can be used in court proceedings, so testers should only submit test reports that are true, accurate, and complete.

Additional information to consider for test reports:

- Test reports should be submitted on approved forms, if required by the *water supplier or administrative authority.*
- Completed forms should be legible, accurate, and should include all relevant or required information.
- Test reports should be completed and signed by the currently certified or licensed backflow assembly tester who performed the test and include a statement certifying the test report is accurate and true.
- Test reports should be provided to the water supplier within a reasonable number of days from the test or as required by the administrative authority. Most suppliers consider receiving the test report within 10 days of the test to be a reasonable time frame.
- Test report should include any repairs that are needed or were made. The backflow assembly tester should *report conditions* that could affect the performance of the assembly, such as any repairs needed or per formed. The tester should notify the property owner of these conditions as well.
- Test reports should include test kit information, such as an identifying serial number and date of the most recent calibration or accuracy verification.
- Test reports should include backflow assembly tester information, such as name, company, address, contact information, and certification or license number, if applicable.

> The Pacific Northwest Section of the American Water Works Association and the Cross-Connection Control Committee profess no expertise in the area of worker safety and related safety regulations.

Figure B-19 is a sample test report form from PNWS-AWWA. It is produced in triplicate to provide copies to the water customer, water supplier and the tester. Forms can be ordered through the PNWS-AWWA website.

Backflow Assembly Tester Ethics

Backflow assembly testers should be aware that their interactions with assembly owners and water suppliers reflect on the backflow prevention industry. Ethical backflow assembly testers should

- Submit complete, accurate test report forms based on the work performed.
- Perform only needed repairs and only when the backflow assembly tester is authorized, qualified, and properly licensed to perform such maintenance tasks.
- Obtain approval from the administrative authority. The tester may need to meet additional licensing requirements to install, remove, replace, or relocate an assembly.
- Apply for permits and inspections from the plumbing or building authority when required.

> The Pacific Northwest Section of the American Water Works Association and its Cross-Connection Control Committee assume no responsibility for any injury or damage to persons or property resulting from the use of this information.

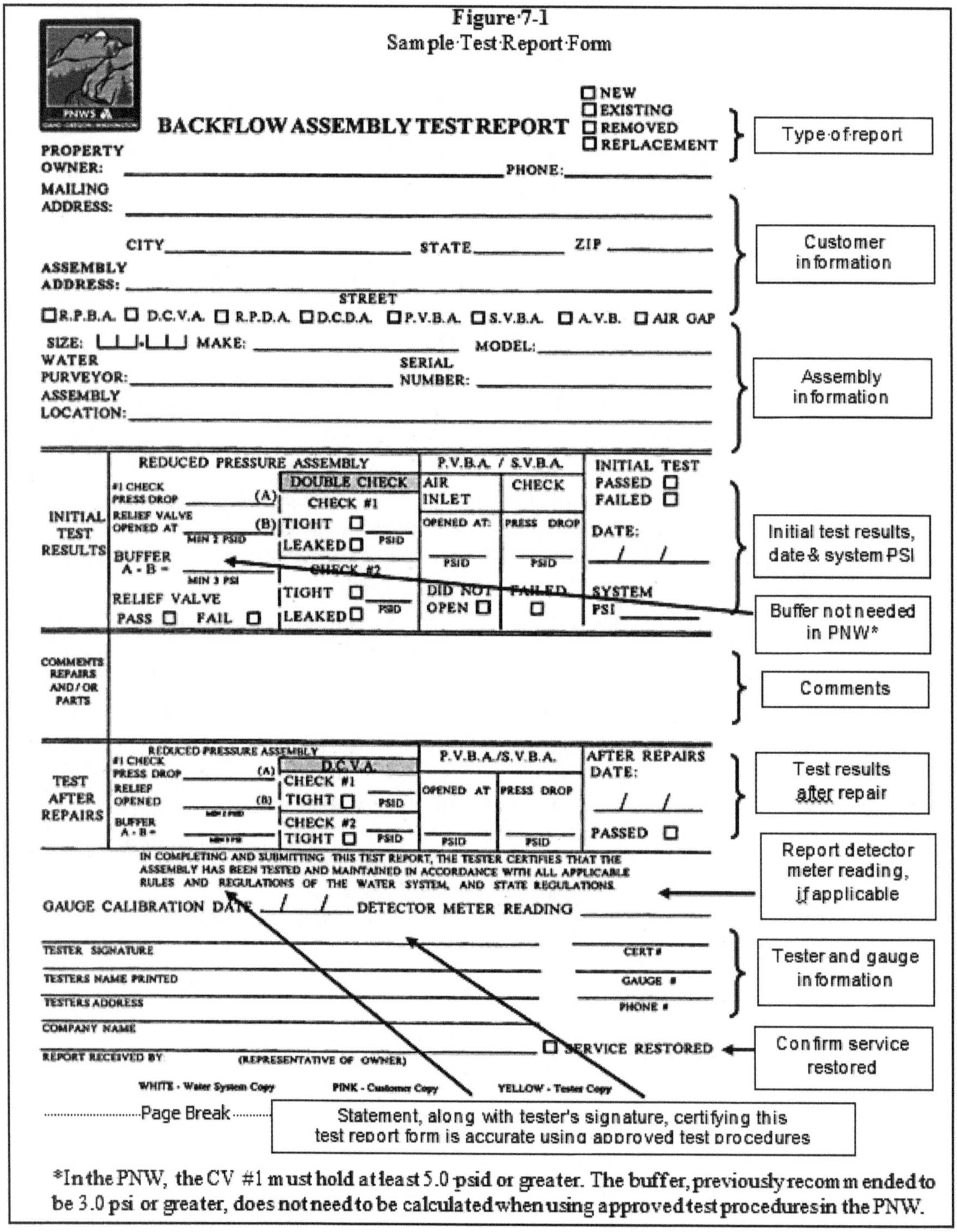
Figure 7-1
Sample Test Report Form

BACKFLOW ASSEMBLY TEST REPORT

☐ NEW ☐ EXISTING ☐ REMOVED ☐ REPLACEMENT

PROPERTY OWNER: ____ PHONE: ____

MAILING ADDRESS: ____

CITY ____ STATE ____ ZIP ____

ASSEMBLY ADDRESS: ____ STREET

☐ R.P.B.A. ☐ D.C.V.A. ☐ R.P.D.A. ☐ D.C.D.A. ☐ P.V.B.A. ☐ S.V.B.A. ☐ A.V.B. ☐ AIR GAP

SIZE: ____ MAKE: ____ MODEL: ____

WATER PURVEYOR: ____ SERIAL NUMBER: ____

ASSEMBLY LOCATION: ____

INITIAL TEST RESULTS: REDUCED PRESSURE ASSEMBLY — #1 CHECK PRESS DROP ____ (A); RELIEF VALVE OPENED AT ____ (B) MIN 2 PSID; BUFFER A - B = ____ MIN 3 PSI; RELIEF VALVE PASS ☐ FAIL ☐. DOUBLE CHECK — CHECK #1 TIGHT ☐ ____ LEAKED ☐ PSID; CHECK #2 TIGHT ☐ ____ LEAKED ☐ PSID. P.V.B.A. / S.V.B.A. — AIR INLET OPENED AT: ____ PSID; CHECK PRESS DROP ____ PSID; DID NOT OPEN ☐; FAILED ☐. INITIAL TEST PASSED ☐ FAILED ☐; DATE: / / ; SYSTEM PSI ____

COMMENTS REPAIRS AND/OR PARTS

TEST AFTER REPAIRS: REDUCED PRESSURE ASSEMBLY — #1 CHECK PRESS DROP ____ (A); RELIEF OPENED ____ (B) MIN 2 PSID; BUFFER A - B = ____ MIN 3 PSI. D.C.V.A. — CHECK #1 TIGHT ☐ PSID; CHECK #2 TIGHT ☐ PSID. P.V.B.A./S.V.B.A. — OPENED AT ____ PSID; PRESS DROP ____ PSID. AFTER REPAIRS DATE: / / ; PASSED ☐

IN COMPLETING AND SUBMITTING THIS TEST REPORT, THE TESTER CERTIFIES THAT THE ASSEMBLY HAS BEEN TESTED AND MAINTAINED IN ACCORDANCE WITH ALL APPLICABLE RULES AND REGULATIONS OF THE WATER SYSTEM, AND STATE REGULATIONS.

GAUGE CALIBRATION DATE / / DETECTOR METER READING ____

TESTER SIGNATURE ____ CERT # ____

TESTERS NAME PRINTED ____ GAUGE # ____

TESTERS ADDRESS ____ PHONE # ____

COMPANY NAME ____

REPORT RECEIVED BY (REPRESENTATIVE OF OWNER) ____ ☐ SERVICE RESTORED

WHITE - Water System Copy PINK - Customer Copy YELLOW - Tester Copy

*In the PNW, the CV #1 must hold at least 5.0 psid or greater. The buffer, previously recommended to be 3.0 psi or greater, does not need to be calculated when using approved test procedures in the PNW.

Figure 8-19 Sample test report form

Backflow assembly testers may need to have current tester certification or licensing from an approved agency or training facility. They may also need to be certified or licensed by multiple agencies to perform the work of testing assemblies. Other business requirements may include appropriate insurance, bond, or other regulatory licenses, such as a contractor or plumber.

Test reports should accurately reflect the condition of the assembly. All repairs should be pre-authorized and noted for the customer, water supplier, and the industry for statistical purposes. Customers and elected officials often challenge testing requirements, or the frequency of testing, by pointing out the lack of failures in testing statistics. *It is critical for the industry to collect accurate data to justify testing requirements or the recommended annual backflow assembly testing frequency.*

Backflow Assembly Tester and Cross-Connection Specialist Safety

In this section, safety procedures and information for backflow assembly testers and cross-connection specialists/inspectors are discussed. The importance of working in a safe manner cannot be over-emphasized.

Water systems and private companies may employ a safety supervisor. They may have their own specific safety procedures that testers and specialists/inspectors must follow. Table 7-1 includes a list of various safety-related publications for further information.

Safety hazards, explanations, and suggestions include the following:

- **Tools:** Tools are manufactured to perform certain tasks. Use of the wrong tool may result in injury to a worker and cause damage to the equipment being repaired. Use the correct tool for the task. When repairing backflow prevention assemblies, consult manufacturer's information or literature for recommendations on proper tools and follow the correct procedures.

> When the *access cover* may also serve as the spring retainer:
>
> *Remove two cover bolts opposite of each other and replace with all-thread rods and nuts.*
>
> *Remove remaining cover bolts, slowly and evenly back off the nuts on the all-thread bolts to release tension and safely remove the cover.*

- **Assembly Springs:** Safety precautions should be followed whenever the cover of a check valve needs to be removed. Refer to manufacturer's information or literature for disassembly and other important information or warnings for the brand, model, and size when performing repairs.

Some check valve modules are held in place by spring pressure. Large assemblies have heavy covers and springs may be uncontained with tension on them. It should never be assumed that springs are contained, even if the manufacturer's literature states they are. Special care should be taken to keep hands and fingers from being pinched or cut while removing or re-assembling check valves.

Before removing a valve cover, close both shutoff valves and release water pressure from the assembly. In reduced-pressure backflow assemblies, the #1 check valve spring is stronger than the #2 check valve spring. This is to create enough of a pressure drop across the check valve for proper operation of the relief valve. To force water through the bypass some detector assemblies use stronger springs than those in the standard assembly of the same make and size. These differences should be considered when estimating spring tension in a check valve.

When a tester or maintenance person needs to remove an assembly, a temporary jumper cable from the downstream piping to the upstream piping should be installed.

On some older assemblies, the **access cover** may also serve as the spring retainer, so use caution when removing the bolts that hold the cover in place. Never remove all bolts holding the cover until verifying there is slack in the

> When a tester or maintenance person needs to remove an assembly, a *temporary jumper cable* from the downstream piping to the upstream piping should be installed.

cover. Two continuous-threaded rods (also called "all thread" rods) and nuts could be used to slowly and evenly back off and remove the covers.

On newer assemblies, the springs are normally a captured mechanism. When bolts on the valve cover are removed, the springs should remain captured within the check valve module.

Remember that all replacement parts should meet the manufacturer's specifications to maintain the assembly's approval status.

- **Assemblies Installed Overhead:** Assemblies installed more than 5 ft above the floor or ground level should have access provided by a permanent platform for the tester or maintenance person. The maximum installation height allowed may vary between jurisdictions. Check with the administrative authority for installation and access requirements. The platform should meet all applicable safety standards and codes.

- **Grounding and Electrical Hazards:** In many jurisdictional areas, it is common practice to ground the electrical system to the water piping system and a ground rod. When a ground rod fails or the electrical system shorts to ground, an electrical charge could energize the water piping.

  During the removal or testing of an assembly, any person touching the assembly or piping could become a better ground than the piping. A person standing in water becomes a better path for an electrical charge to travel through them as it discharges to the ground. Figure B-4 illustrates the permanent installation of ground wire installed around an assembly. When a tester or maintenance person needs to remove an assembly, a *temporary jumper cable* from the downstream piping to the upstream piping should be installed. (Figure B-20.)

When testing an assembly, care should be taken to avoid splashing or spraying water onto nearby electrical equipment that could result in damage to the equipment or an electric shock to the tester or maintenance person.

- **Thrust Restraint:** Some assemblies may be connected to adjoining piping in such a manner that places them under tension or compression. The thrust (force) is caused by water pressure within piping acting against a closed valve. Piping must have thrust restraints installed to prevent pipes from separating.

  Thrust restraints are often buried and not always evident. Injury or property damage could result if an assembly without thrust restraint is removed.

  It is safer to assume there are no thrust restraints when replacing or working on larger assemblies in vaults. A tester or maintenance person should block or add restraints, as required, to prevent piping from separating

- **Hazardous Materials:** Hazardous and toxic materials may be present near a backflow assembly, as this is often the reason why an assembly is needed. Proper caution should be used when working in such an area. Respirators, gloves, protective coveralls, and other personal protective equipment may be required. Be aware of posted warnings and other indications of hazards such as an eye wash or emergency shower stations and material safety data sheets.

- **Confined Space Entry:** *Confined space entry* training may be required and procedures should be followed when testing or surveying backflow assemblies located in underground vaults or other areas that meet the definition of a confined space. A confined space is defined as any space that

  — Has a limited means of entry or exit.

  — Is not intended for continuous occupancy.

  — Is large enough to bodily enter.

> *Personal protective equipment* may be needed when working with hazardous materials. Examples are listed in the workplace safety section.

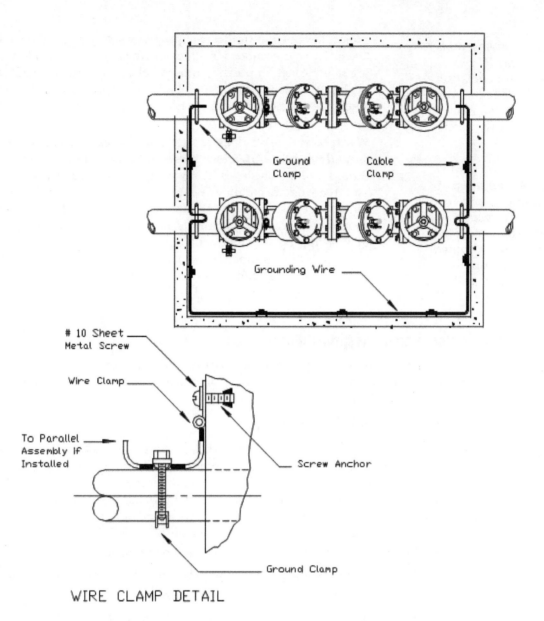

Figure 8-20   Ground wire installation

Confined spaces can be subject to an accumulation of toxic or flammable contaminants, engulfment, or an oxygen-deficient atmosphere. Equipment such as air quality monitors, ventilators, retrieval tripods, and harnesses may be required in some jurisdictions and private facilities. Some confined spaces may require the use of an attendant. No one should attempt to rescue someone in a confined space without proper training and equipment. Would-be rescuers are commonly inexperienced and untrained. The majority of confined space accidents result from rescue attempts.

- **Traffic:** Assemblies may be installed in locations near roadways, parking areas, or sidewalks. A tester or maintenance person may need to park a vehicle near a roadside. Both these situations are

> *Additional requirements, training and equipment may be needed for confined space entry.*

> *Backflow Assembly Testers, Cross-Connection Specialists and maintenance personnel* should take responsibility for their own safety.

> *Only trained and properly equipped personnel should ever attempt a rescue in a confined space. Many times the untrained rescuer ends up needing to be rescued.*

subject to vehicular traffic, so appropriate safety precautions should be taken. These precautions include wearing a reflective safety vest, placing safety cones, or utilizing other work area protection such as warning lights, high level caution signs, flares, or providing a flag person or emergency personnel to direct traffic for the safety of the worker and the general public.

- **Workplace Safety:** *Backflow assembly testers, cross-connection specialists or maintenance personnel* may work in a variety of conditions and on job sites with a variety of safety concerns. It is important that workers maintain awareness of their surrounding work area for any specific or potential hazards. Personal protective equipment should be worn, such as a hard hat, safety glasses, safety goggles, gloves, appropriate foot protection, ear protection, and protective clothing. Often companies post information about hazards at their own sites or they employ safety personnel to oversee work at their facilities.

Water Supplier Responsibilities

One element listed for a cross-connection control program is to establish a quality assurance and quality control program for providing oversight of testers working within the supplier's water distribution system.

Backflow Assembly Test Procedures

To ensure that test results obtained from certified or licensed backflow assembly testers are reliable, approved testing procedures should be used. Approved test procedures may be specified by the state or regulatory authority or the water supplier.

The following are publications or sources that provide test procedures for reduced-pressure backflow assemblies, double check valve assemblies, pressure vacuum breaker, and spill-resistant vacuum breaker assemblies (Table B-1.):

- Manufacturer's literature included with test equipment
- Manufacturer's literature included with backflow prevention assemblies
- New England Water Works Section of AWWA
- *Cross-Connection Control Manual,* USEPA Publication 816-R-03-002
- American Backflow Prevention Association (ABPA)
- *Backflow Prevention Assemblies—Series 5000,* American Society of Sanitary Engineering
- *Manual of Cross-Connection Control,* 10th ed., Foundation for Cross-Connection Control and Hydraulic Research, University of Southern California
- *Backflow Prevention Theory and Practice,* University of Florida TREEO Center

Test procedures listed above may vary with the type of test equipment used, the type of test performed to determine the operating performance of the assembly, and the criteria used to determine compliance with the operating requirements. Some tests may vary only in the sequence of the steps or in the method of troubleshooting or diagnosing problems.

Table 8-1 Safety-related publications

Publisher	Publication or Notes
NIOSH (National Institute for Occupational Safety and Health) www.cdc.gov.niosh (800) 232-4636	"Working in Confined Spaces," Publication #80-106 "A Guide to Safety in Confined Spaces", Publication #87-113
USDOL Occupational Safety and Health Administration www.osha.gov (800) 321-6742	Title 29 Code of Federal Regulations "Atmospheric Testing in Confined Spaces"
USDOL Region 10 Seattle, WA (206) 553-5930	WA, OR and AK operate OSHA programs under a plan approved and monitored by USDOL
State of Oregon Occupational Safety and Health Administration www.orosha.org (800) 922-2689	"Small Business Handbook", Pub. 2209-02R "Personal Protective Equipment", Pub. 3151-12R
State of Washington Department of Labor and Industries Dept. of Occupational Safety and Health www.lni.wa.gov/safety (800) 423-7233	"Right to Know for Small Business" "Understanding Right to Know" Chemical Hazard Communication Guidelines
State of Idaho Boise, ID 83706 (208) 321-2960	See USDOL OSHA or Region 10 or Consultation Project Offices http://oshcon.boisestate.edu/

This list of safety publications provides only some of the available information.

Recommended test procedures are those that are based on evaluating the operating requirements specified in the design standard for an assembly using a differential pressure test kit. The important operating requirements for assemblies, such as those established in AWWA standards, are summarized in Appendix Q.

Detailed step-by-step procedures for testing backflow prevention assemblies are not provided in this manual. Backflow assembly testers should receive training through an approved backflow assembly tester course to become familiar with the testing procedures. These include flushing test cocks, installing test cock adapters, isolating the assembly by closing shutoff valves (as required by approved procedures), connecting test equipment, bleeding air from the test equipment, and other common tasks.

> Verify with the *administrative authority* which test procedures are approved.

In the place of step-by-step procedures, this manual provides an outline of the recommended test objective, a summary of the recommended test method, and the minimum reporting requirements for each type of assembly. The purpose of this information is to provide the general recommended minimum performance requirements when testing various types of assemblies.

Alternate test procedures may be accepted by some jurisdictions. Common alternate test procedures may include the use of a sight tube, a duplex pressure gauge, or testing an RPBA in the direction of flow. The backflow assembly tester should verify with the administrative authority which *test procedures are approved* and whether alternate test procedures are accepted.

> *This overview of test procedures is to assist the Cross-Connection Specialist in understanding how testing is performed.*

Preliminary Steps to Testing a Backflow Assembly

Prior to initiating a test of any backflow prevention assembly, the following procedures should be followed:

1. Obtain permission from the owner, or their representative, to shutdown the water supply. Since all testing is accomplished under no-flow conditions, the owner needs to be aware that the water supply will be temporarily shutoff while testing is being performed. Some commercial and industrial operations require constant and uninterrupted water supplies for cooling, boiler feed, seal pump water, or other such uses and water service interruptions cannot be tolerated. The water supply to hospitals and continuous process industries cannot be shutoff without planned and coordinated shutdowns.

    For premise isolation assemblies, although notice can be given by the supplier for an interruption of service, whenever possible, it is preferable to cooperate with the owner to arrange a mutually-agreeable time for a shutdown.

    The request to shutoff the water supply is a necessary prerequisite to protect the customer as well as limit the liability of the tester.

    Concurrent with the request for permission to shutoff the water, it is advisable to point out to the owner that while the water is shutoff during the test period, any inadvertent use of water within the building will reduce the water pressure to zero. Backsiphonage could result in the building's plumbing system being contaminated through cross connections.

    To address this situation, it is recommended that the owner caution the occupants of the building not to use water until the backflow assembly test is completed and the water pressure restored. Additional options available to the owner would be the installation of two backflow assemblies in parallel that would enable a protected by-pass flow around the assembly to be tested.

2. Determine the type of assembly to be tested (e.g., RPBA, DCVA, PVBA, or SVBA).

3. Determine the direction of flow (reference directional flow arrows, the location of test cocks, or wording provided by the manufacturer on the assembly).

4. Number the test cocks (mentally), flush them of potential debris (in the correct order, if specified by approved procedures), and assemble appropriate test cock adapters and bushings that may be required.

5. Close shutoff valve #2, the downstream isolating valve (some test procedures delay this step until later).

6. Attach the test equipment in the manner appropriate to the assembly being tested and the specific test being performed.

> Testing the RPBA requires closing shut-off valve #2. After closing the shut-off valve, if the relief valve vents, report that check valve #1 leaks and end the test.

> When testing an RPBA, the differential pressure being measured by the test gauge is the difference of pressure between the supply water and the zone of reduced pressure. This holds true for every component tested on an RPBA.

> The abbreviation *psid* refers to differential pressure.

Introduction to Testing Each Type of Backflow Assembly

The information provided in this section is designed to help water suppliers understand how each component of an assembly performs. A very basic description of the testing procedure is also explained to give some guidance to a water supplier overseeing quality assurance/quality control of their testing program.

The following backflow testing information is commonly used in the Pacific Northwest and is organized by assembly component. This is for general information only and may not be the test procedures accepted by all *administrative authorities.*

Test Procedures for Reduced-Pressure Backflow and Reduced Pressure Detector-Type Assemblies using a Differential Pressure Gauge Test Kit

Relief Valve (RPBA/RPDA)

Performance Criteria: During normal operating conditions, whether or not there is flow through the assembly, the pressure in the zone between the check valves (zone of reduced pressure) shall be at least 2 psi less than the pressure on the supply (inlet) side of the assembly. When there is no flow from the supply side of the assembly and the supply pressure drops to 2 psi, the pressure within the zone shall be atmospheric. As the supply pressure drops below 2 psi, the relief valve shall continue to open and shall reach and maintain the fully open position as the supply pressure drops to atmospheric or lower. [AWWA C511 Sec. 4.3.2.1, 4.3.2.2]

Test Objective, Method, and Reporting Requirements: The first test objective is to determine the initial opening point and operation of the differential pressure relief valve by increasing the pressure in the area between the two check valves (the zone of reduced pressure). This is accomplished by

- Introducing the higher supply pressure through the differential pressure gauge test kit, into the lower pressure of the zone of reduced pressure.
- The first drop of water observed is the initial opening point of the relief valve.
- Note the reading on the differential pressure gauge (must be 2.0 *psid* or greater to pass).

Record this reading on the test report form as the initial opening point of the Relief Valve.

NOTE: *Some test procedures also include verification that the relief valve will continue to open as the* **differential pressure** *drops below the relief valve opening point.*

Check Valve #2 (RPBA/RPDA)

Performance Criteria: Check valve #2 will hold tight against backpressure.

Test Objective, Method, and Reporting Requirements: check valve #2 is verified for holding tight by introducing backpressure on the downstream side of the check valve. This is accomplished by

- Introducing higher supply pressure through the differential pressure gauge test kit into the downstream side of check valve #2.
- Observing the differential pressure gauge. The psid may or may not decrease. If the relief valve remains closed and the gauge is stabilized, check valve #2 will be recorded as holding tight.

Record "tight" or "leaked" on the test report form for check valve #2. (No psid value is entered on the test report form for Check Valve #2.)

Check Valve #1 (RPBA/RPDA)

Performance Criteria: The static psid across check valve #1 will be greater than initial relief valve opening point and at a minimum of **5.0** *psid*.

> The **5.0 psid** across check valve #1 eliminates past recommendations for a 3.0 psid buffer above the relief valve opening point. A check valve #1 holding value of 5.0 psid will prevent backflow, but will discharge more often due to supply pressure fluctuations.

> *Some test procedures for RP-type assemblies may require verification of the distance of the air gap below the Relief Valve discharge vent.*

Test Objective, Method, and Reporting Requirements: (Note: This test is performed after testing check valve #2.) (Table B-2.)

To test check valve #1 for tightness in the direction of flow and determine the static psid across check valve #1. This is accomplished by:

- Maintaining backpressure on the downstream side of check valve #2.
- Create flow through the assembly using the differential pressure gauge and then return to a static flow condition.
- When the gauge stabilizes at 5.0 psid or greater and the relief valve remains closed, check valve #1 meets the performance criteria.

Record "tight" or "leaked" on the test report form for check valve #1 and enter the observed psid value.

Bypass Meter on RPDA

Performance Criteria: The bypass meter should register any flow (e.g., 3–5 gal) that occurs through the assembly (mainline or bypass). However, it is not necessary that the meter accurately register the flow.

Test Objective, Method and Reporting Requirements: Partially open the mainline assembly's test cock #4. Observe the bypass meter; the meter dial should move to register flow.

In addition, if test cock #4 of the mainline assembly is located on the bypass piping (rather than on the body of the main line assembly), close shutoff Valve #2 on the bypass assembly and partially open test cock #4. If flow continues from the test cock, this indicates the bypass connection to the body of the mainline assembly is not restricted.

Record on the test report form that the "detector" meter registered flow (if required by the administrative authority).

Observe and verify that the RPBA/RPDA

- Is designated as an approved assembly by the administrative authority.
- Is installed in the approved orientation.
- Provides the correct protection for the potential hazard (the water supplier has the responsibility to verify the proper type of assembly was installed for protection from the degree of hazard).
- Is correctly installed with approved clearances for testing and maintenance.
- Provides adequate drainage.
- Test results are properly documented on a test report form:
  — Relief Valve opening: initial opening 2.0 psid or greater to pass
  — Check Valve #2: holds tight (no psid value)
  — Check Valve #1: holds tight at 5.0 psid or greater to pass
  — Detector meter: registers flow and the bypass meter reading is recorded (if required)

Table 8-2 RPBA/RPDA test reporting

Relief valve [≥2.0 psid]	Dripped at: or failed to open? ____ (check) Continued to open?	__ __ . __ psi yes ____, no ____
Check valve #2 [Holds tight]	Pressure drop: Valve tight against backpressure?	 yes ____, no ____
Check valve #1 [5.0 psid]	Pressure drop: valve tight?	__ __ . __ psi yes ____, no ____
Air gap distance approved?		yes ____, no ____
Test cock #4 opened, meter moved?		yes ____, no ____
Detector meter reading: _____		

Test Procedures for Double Check Valve and Double Check Detector-type Assemblies using a Differential Pressure Gauge Test Kit

Check Valve #1 and Check Valve #2

Performance Criteria: The check valves shall be loaded internally so that when the supply pressure is at least 1 psi and the outlet pressure is atmospheric, each check valve will be drip-tight in the normal direction of flow. [AWWA C510 Sec. 4.3.2.1]

Test Objective, Method and Reporting Requirements: To test check valve #1 and check valve #2 for tightness in the direction of flow and determine the static pressure drop across each check valve using a differential pressure gauge test kit. This is accomplished by

- Connecting the differential pressure gauge to the upstream side of the check valve being tested. The test gauge center line must be held at the same level as the downstream test cock reference point (outlet of the test cock or the water level if a sight tube is used).

- Closing both shut-off valves on the DCVA to isolate the pressure within the assembly.

- Releasing water downstream of the check valve so the pressure is reduced to atmospheric.

- When flow from the downstream test cock ceases and the differential pressure gauge reading has stabilized, record the psid across the check valve.

- The check valve holding tight at a minimum of 1.0 psid or greater.

Record "tight" or "leaked" on the test report form for check valve #1 or check valve #2 along with the psid values for each check valve. Note: leaked indicates the gauge reading dropped to 0.

> When testing a DCVA, the *differential pressure gauge reading* being measured is the difference between the supply pressure on the upstream side of the check valve and the atmospheric pressure on the downstream side of the check valve.

> When water flow stops from the downstream check valve test cock and the gauge reading has *stabilized*, record the check valve as closed tight at the psid reading on the gauge.

Bypass Meter on DCDA

Performance Criteria: The bypass meter should register any flow (e.g., 3 to 5 gallons) that occurs through the assembly (mainline or bypass). However, it is not necessary that the meter accurately register the flow.

Test Objective, Method and Reporting Requirements: Partially open the mainline assembly's Test Cock #4. Observe the bypass meter; meter dial should move to register flow.

In addition, if test cock #4 of the mainline assembly is located on the bypass piping (rather than on the body of the main line assembly), close shutoff valve #2 on the bypass assembly and partially open test cock #4. If flow continues from the test cock, this indicates the bypass connection to the body of the mainline assembly is not restricted.

Record on the test report form that the "detector" meter registered flow (if required by the administrative authority). (Table B-3)

Observe and verify that the DCVA/DCDA

- Is designated as an approved assembly by the administrative authority.
- Is installed in the approved orientation.
- Provides the correct protection for the potential hazard (the water supplier has the responsibility to verify the proper assembly was installed for protection from the degree of hazard).
- Is correctly installed with approved clearances for testing and maintenance.
- Test results are properly documented on a test report form:
  — Check valve #1: holds tight at 1.0 psid or greater
  — Check valve #2: holds tight at 1.0 psid or greater
  — Detector meter: registers flow and record the bypass meter reading (if required)

Test Procedures for Pressure Vacuum Breaker Assemblies using a Differential Pressure Gauge Test Kit
air inlet valve (PVBA)

Performance Criteria: The air inlet valve will initially open when the differential pressure in the body is no less than 1.0 psi above atmospheric pressure. The air inlet valve will also continue to open fully when water has drained from the body of the assembly.

Table 8-3 DCVA/DCDA test reporting

Check valve #1 [≥1.0 psid]	Pressure drop: Valve tight?	__ __ . __ psi yes ____, no ____
Check valve #2 [≥1.0 psid]	Pressure drop: Valve tight?	__ __ . __ psi yes ____, no ____
Test cock #4 opened, metered moved? Detector meter reading: _____		yes ____, no ____

Test Objective, Method and Reporting Requirements: Determine the initial opening point of the air inlet valve using a differential pressure gauge test kit. This is accomplished by:

- Connecting the differential pressure gauge to the upstream side of the air inlet valve and holding the gauge at the centerline of the air inlet valve.
- Closing both shut-off valves on the PVBA to isolate pressure within the assembly.
- Slowly draining water pressure from the assembly.
- Observing the differential pressure gauge reading at the initial opening point of the air inlet valve.
- The initial opening point must be 1.0 psid or greater to pass.
- Closing the test kit valve and removing the hose from the test cock. As the assembly is drained through the open test cock, the air inlet valve should continue to open.
- When water has drained from the assembly, the air inlet must open fully.

Record on the test report form the initial opening psid value of the air inlet valve. Observe or note that the air inlet valve continued to fully open.

Check Valve (PVBA)

Performance Criteria: The check valve is internally loaded and drip tight in the normal direction of flow with 1.0 psid.

Test Objective, Method, and Reporting Requirements: Test the check valve for tightness in the direction of flow and determine the static pressure drop across the check valve using a differential pressure gauge test kit. This is accomplished by

- Connecting the differential pressure gauge to the upstream side of the check valve and holding the gauge at the centerline of test cock #2.
- Closing both shut-off valves on the PVBA to isolate the pressure within the assembly.
- Reducing water downstream of the check valve to atmospheric.
- Observing the differential pressure gauge reading.
- When water stops flowing from the assembly and the gauge reading stabilizes, the check valve should close tight at 1.0 psid or greater.

Record the differential pressure gauge reading on the test report form as the check valve pressure drop.

Observe and verify that the PVBA

- Is designated as an approved assembly by the administrative authority.
- Is properly installed 12 in. above the highest downstream outlet.
- Provides the correct protection for the potential hazard (the water supplier has the responsibility to verify the proper assembly was installed for protection from the degree of hazard).
- Is correctly installed with approved clearances for testing and maintenance.
- Test results are properly documented on a test report form:
  — Air inlet valve initially opens at 1.0 psid or greater and then continues to open fully.
  — Check valve holds tight at 1.0 psid or greater.

Test Procedure for Spill-Resistant Vacuum Breaker Assemblies using a Differential Pressure Gauge Test Kit
Check Valve and air inlet valve (SVBA)

Performance Criteria: The check valve is internally loaded and drip tight in the direction of flow with 1.0 psid. The air inlet valve should initially open at 1.0 psid or greater and continue to open fully.

Test Objective, Method, and Reporting Requirements for Check Valve and Air Inlet Valve:

Note: Once the test gauge is elevated when testing an SVBA, it should remain elevated while testing the check valve and air inlet valve opening point.

Test the check valve for tightness in the direction of flow and determine the static pressure drop across the check valve using a differential pressure gauge test kit. Determine the initial opening point of the air inlet valve using a differential pressure gauge test kit. This is accomplished by

Check Valve Testing
- Connecting the differential pressure gauge to the upstream side of the check valve (there is only one test cock on this assembly).
- Holding the center line of test gauge at the vent valve outlet (vent screw) while testing both the check valve and air inlet valve.
- Closing the shut-off valves on the SVBA to isolate the pressure within the assembly.
- Reducing water downstream of the check valve to atmosphere.
- Observing the differential pressure gauge reading when flow from the vent valve ceases and the gauge reading has stabilized.
- The check valve should hold tight at 1.0 psid or greater.
- The test gauge continues to remain elevated for the air inlet valve test.

Record the differential pressure gauge reading on the test report form as the Check Valve pressure drop.

Air Inlet Valve Testing
- With the test gauge still elevated, reduce pressure on the upstream side of the check valve by draining water from assembly through the differential pressure gauge.
- The air inlet valve should initially open at 1.0 psid or greater.
- Record the initial opening point of air inlet valve.
- With test kit lowered, close the high bleed valve and remove the test gauge hose from the test cock. When flow ceases from test cock, observe that the air inlet valve continued to fully open.

Record the initial opening psid of the air inlet valve. Observe or note that the air inlet valve continued to open fully. (Table B-4.)

Observe and verify that the SVBA
- Is designated as an approved assembly by the administrative authority.
- Is properly installed 12 in. above the highest downstream outlet.

Table 8-4   PVBA/SVBA test reporting

Check valve #1 [≥1.0 psid]	Pressure drop: Valve tight?	___.__ psi yes ____, no ____
Air inlet [≥1.0 psid]	Opened at: Air inlet fully opened	___.__ psi yes ____, no ____

Table 8-5   Approved minimum test result values

Assembly Type ▼	◄ ASSEMBLY COMPONENT ►			
	Check Valve #1 Holds at:	Check Valve #2 Holds at:	Relief Valve Opens at:	Air Inlet Valve Opens at:
RPBA/RPDA	Tight ≥ 5.0 psid	Tight (no psid determined)	≥ 2.0 psid	n/a
DCVA/DCDA	≥ 1.0 psid	≥ 1.0 psid	n/a	n/a
PVBA	≥ 1.0 psid	n/a	n/a	≥ 1.0 psid and fully open
SVBA	≥ 1.0 psid	n/a	n/a	≥ 1.0 psid and fully open

NOTE: n/a indicates the assembly does not have that component.

- Provides the correct protection for the potential hazard (the water supplier has the responsibility to verify proper assembly was installed for protection from the degree of hazard).
- Is correctly installed with approved clearances for testing and maintenance.
- Test results are properly documented on the test report form:
  — Check valve holds tight at 1.0 psid or greater.

Air inlet valve initially opens at 1.0 psid or greater and continues to fully open.

In the Pacific Northwest, approved test procedures use a differential pressure gauge to obtain test results for each component of an assembly. The values recorded must be greater than or equal to the minimum values shown in Table B-5. All components of an assembly must meet or exceed these values to pass the performance test.

> *Verify with the administrative authority whether a type of test kit or specific manufacturer and model of test equipment requires approval.*

Backflow Assembly Test Equipment

Backflow assembly test procedures are performed using test equipment commonly called "test kits." Administrative authorities may specify which type of test kit is approved for testing assemblies such as a duplex or differential pressure gauge. Manufacturers produce test kits that can be analog (Figure B-3) or electronic/digital (Figure B-4). The most common type of test kit approved for use in testing backflow prevention assemblies is the differential pressure gauge. This test kit comes in several models:

- Two valve
- Three valve
- Five valve

Some agencies may grant approval of specific test equipment manufacturers and models. Examples of test equipment manufacturers include:

Acugauge   Gage-It

> In the Pacific Northwest, differential pressure gauge test kits are approved for testing. Annual accuracy verification of test kits is required.

Ames   Mid-West
Barton   Orange Research
Conbraco   Pro-Master (Astra)
Duke Products Watts
Febco   Wilkins

Recommended Maintenance for Test Equipment

A test kit used to test backflow prevention assemblies should be periodically verified for accuracy. This verification should occur at least annually, or more often if damage occurs or false readings are suspected. Calibration may be required to adjust the test kit using established standards traceable to the National Institute of Standards and Technology (NIST).

The following recommendations can help prolong the use and accuracy of the test kit:

- Fully flush assembly test cocks to dislodge any debris before attaching gauge hoses to avoid pulling debris into the test kit.
- If possible, use in-line filters on each test kit hose, flush filters regularly, and replace them as needed.
- Drain the test kit between uses to relieve the internal pressure and to help prevent the possibility of freeze damage or corrosion.
- Leave control valves open to prevent damage to seats.
- Store the test kit where it will not be exposed to temperature extremes or moisture.
- Store and transport the test kit in a protective case or tool box.
- Keep fittings, tools, and other equipment separate from the test kit to avoid damage in transit.
- Inspect hoses, knobs, valves, faceplate seals, or other components for wear or damage and replace as needed.

Differential Pressure Gauge Test Equipment Description

- Differential pressure gauge reads from 0 to 15 psid in 0.1 or 0.2 psid graduations.
- Accurate to plus or minus 0.2 psid.
- Three high pressure hoses in approximately 5–6 ft lengths with minimum ¼-in. inside diameter (ID) and screw-type couplings on each hose end.
- ¼-in. needle valves for fine control of flows.
- Connects to a backflow assembly's test cocks for testing purposes. Appropriate adapter fittings may be needed.

Adapter fittings are sized at ¼-in., ½-in., and ¾-in. to allow test equipment to be connected to various sizes of test cocks located on backflow assemblies.

Figures B-21 through B-27 illustrate the locations where test equipment hoses connect to the test cocks of each type of assembly, and the condition of the internal components when they are under test.

TESTING PROCEDURESOR METHODS 139

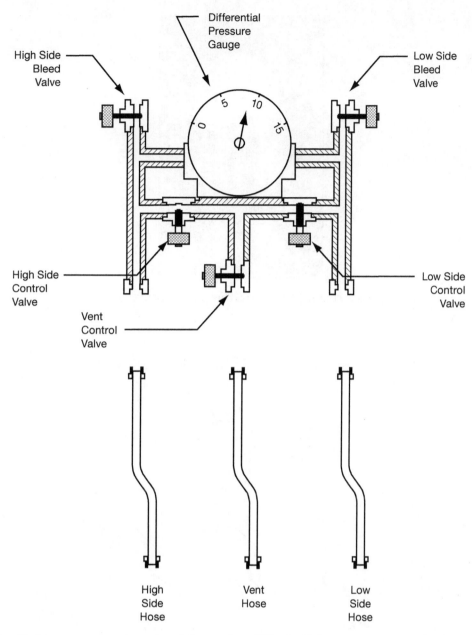

Figure 8-21 Major component parts of five-valve differential pressure gauge test equipment

140 BACKFLOW PREVENTION AND CROSS-CONNECTION CONTROL

Figure 8-22 Differential pressure gauge test kit, five-valve model

(Photo courtesy of Wilkins and BMI.)

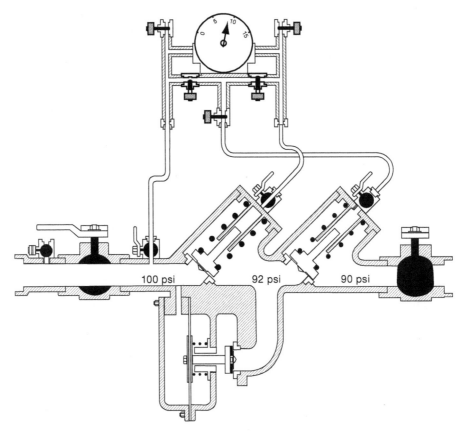

Figure 8-23 Differential pressure gauge showing hose connections to test the components of an RPBA

TESTING PROCEDURES OR METHODS 141

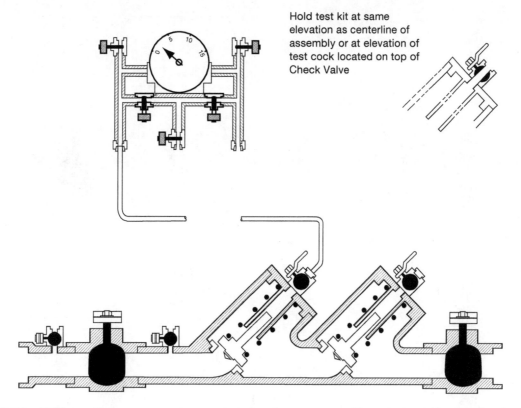

Figure 8-24   Differential pressure gauge showing hose connections to test the #1 check valve of a DCVA

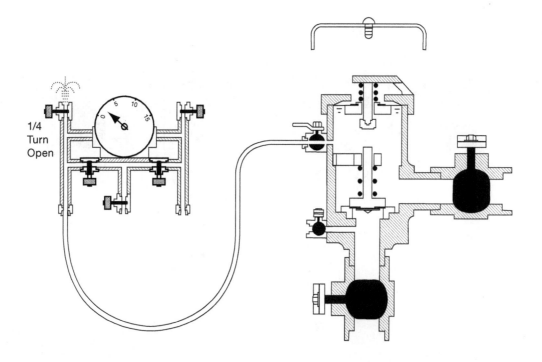

Figure 8-25   Differential pressure gauge showing hose connections to test a PVBA air inlet

AWWA Manual M14

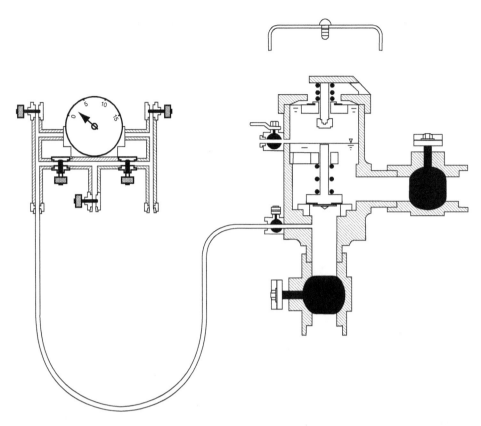

Figure 8-26   Differential pressure gauge showing hose connections to test a PVBA check valve

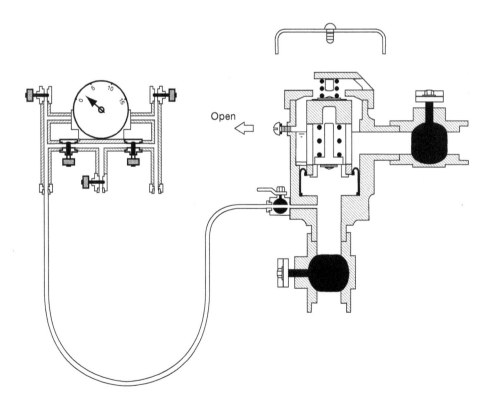

Figure 8-27   Differential pressure gauge showing hose connections to test both the check valve and air inlet of an SVBA

INFORMATION PROVIDED BY THE AMERICAN SOCIETY OF SANITARY ENGINEERING (ASSE) FIELD TEST PROCEDURES USING THREE- AND FIVE- VALVE TEST KITS

Three-Valve Test on an ASSE 1013

Reduced Pressure Principle Backflow Preventers (RP) and Reduced-Pressure Principle Fire Protection Backflow Preventers (RPF)

Administrative Issues

A. Initial arrangements shall be made with the responsible party to schedule the test.
B. Arrangements/notifications shall be made where a continuous water supply is necessary or where testing creates a special hazard, inconvenience, or risk in buildings or for piping systems. Examples of such buildings or systems include, but are not limited by enumeration, to hospitals, manufacturing plants, fire protection systems, alarm companies/fire departments, insurance carriers, and others where special notification is necessary.
C. The proper information, reporting forms, and equipment shall be gathered to properly conduct the test.

Site Issues

A. Safety Inspection

The initial field evaluation shall be made for safety hazards in accordance with applicable federal, state, and local safety regulations and statutes. Some of the issues to be examined are, but are not limited by enumeration, to confined spaces, ventilation, access, oxygen content, chemical, electrical, or flammable hazards, hazards related to elevation of devices, hazards to the tester, and other persons.

B. Inspection of the Installation

The proper application of the assembly shall be confirmed for code compliance with respect to the degree of hazard, markings, prohibited locations (i.e., where subjected to flooding or freezing), and special installation requirements. The assembly orientation and direction of flow shall be confirmed as proper. The assembly shall be checked for alterations or special needs such as, but not limited by enumeration, to the adequacy of the air gap, the evidence of illegal bypasses, and the adequacy of drainage systems from the assembly. The general appearance of the device shall be checked for condition of excessive discharge, condition of the shutoff valves, test cocks, relief valve, air gap, and adequacy of drainage should leakage occur.

Field Testing Requirements

Flush Test Cocks (TC)

Step #1—Install test adapters (if applicable)
Step #2—Open TC #4—let trickle
Step #3—Open TC #1—close
Step #4—Open TC #2—close
Step #5—Open TC #3—close
Step #6—Close TC #4

Attach Test Kit

Step #1—Close high and low valves and open bypass valve on test kit
Step #2—Attach high hose to TC #2
Step #3—Attach low hose to TC #3
Step #4—Open TC #2
Step #5—Open high valve—bleed air—close
Step #6—Open TC #3

Step #7—Open low valve—bleed air—close
Step #8—Attach bypass hose to TC #4
Step #9—Open high valve
Step #10—Loosen bypass hose at TC #4- bleed air- tighten

Test #1—Tightness of the #2 Shutoff Valve

Step #1—Close #2 shutoff valve
Step #2—Open TC #4
Step #3—Close TC #2
Step #4—Read differential gauge

- Test Results -

If the differential gauge reading remains steady, record #2 shutoff valve as tight.

Test #2—#2 Check w/Back Pressure Test

If in Test #1, the differential gauge remains steady, record #2 check valve as tight.

Test #3—#1 Check Differential

Step #1—Close TC #4
Step #2—Close high valve
Step #3—Remove bypass hose from TC #4
Step #4—Open TC #2
Step #5—Open low valve to cause differential gauge reading to increase
Step #6—Close low valve

- Test Results -

Record gauge value.

Record the pressure differential across #1 check valve.

If the differential reading is 5 psid or above, record #1 check valve as tight.

Test #4—Relief Valve Opening

Step #1—Close bypass valve
Step #2—Open high valve
Step #3—Slowly open low valve

- Test Results -

Record relief valve psid opening point. The relief valve must drip when the differential gauge is 2 psid or above.

Restore System

Step #1—Close all TCs
Step #2—Remove hoses
Step #3—Open all valves on test kit to drain water
Step #4—Restore #2 shutoff valve to pretest state

The tester shall provide copies of the test results to the owner and other appropriate parties as required. The tester shall maintain a copy for his/her records.

Five-Valve Test on an ASSE 1013

Reduced-Pressure Principle Backflow Preventers (RP) and Reduced-Pressure Principle Fire Protection Backflow Preventers (RPF)

Administrative Issues

A. Initial arrangements shall be made with the responsible party to schedule the test.
B. Arrangements/notifications shall be made where a continuous water supply is necessary or where testing creates a special hazard, inconvenience, or risk in buildings or for piping systems. Examples of such buildings or systems include, but are not limited by enumeration, to hospitals, manufacturing plants, fire protection systems, alarm companies/fire departments, insurance carriers, and others where special notification is necessary.
C. The proper information, reporting forms, and equipment shall be gathered to properly conduct the test.

Site Issues

A. Safety Inspection

The initial field evaluation shall be made for safety hazards in accordance with applicable federal, state, and local safety regulations and statutes. Some of the issues to be examined are, but are not limited by enumeration, to confined spaces, ventilation, access, oxygen content, chemical, electrical, or flammable hazards, hazards related to elevation of devices, hazards to the tester, and other persons.

B. Inspection of the Installation

The proper application of the assembly shall be confirmed for code compliance with respect to the degree of hazard, markings, prohibited locations (i.e., where subjected to flooding or freezing), and special installation requirements. The assembly orientation and direction of flow shall be confirmed as proper. The assembly shall be checked for alterations or special needs such as, but not limited by enumeration, to the adequacy of the air gap, the evidence of illegal bypasses, and the adequacy of drainage systems from the assembly. The general appearance of the device shall be checked for evidence of excessive discharge, condition of shutoff valves, test cocks, relief valve, air gap, and adequacy of drainage should leakage occur.

Field Testing Requirements

Flush Test Cocks (TC)

Step #1—Install test adapters (if applicable)
Step #2—Open TC #4—let trickle
Step #3—Open TC #1—close
Step #4—Open TC #2—close
Step #5—Open TC #3—close
Step #6—Close TC #4

Attach Test Kit

Step #1—Close high, low, and bypass valves and high and low bleed valves on test kit
Step #2—Attach high hose to TC #2
Step #3—Attach low hose to TC #3
Step #4—Open TC #2
Step #5—Open high bleed valve—bleed air—close
Step #6—Open TC #3
Step #7—Open low bleed valve—bleed air—close
Step #8—Attach bypass hose to TC #4
Step #9—Open high valve

Step #10—Open bypass valve
Step #11—Loosen bypass hose at TC #4—bleed air—tighten
Step #12—Slowly open low bleed valve to cause differential reading to rise—close

Test #1—Tightness of #2 Shutoff Valve
Step #1—Close #2 shutoff valve
Step #2—Open TC #4
Step #3—Close TC #2
Step #4—Read differential gauge

- Test Results -

If the differential gauge reading remains steady, record #2 shutoff valve as tight.

Test #2—#2 Check w/Back Pressure Test

If in Test #1, the differential gauge remains steady, record #2 check valve as tight.

Test #3—#1 Check Differential
Step #1—Close bypass valve
Step #2—Open TC #2
Step #3—Open low bleed valve to cause reading to increase
Step #4—Close low bleed valve

- Test Results -

Record gauge value.

Record the pressure differential across #1 check valve.

If the differential reading is 5 psid or above, record #1 check valve as tight.

Test #4—Relief Valve Opening
Step #1—Slowly open low valve

- Test Results -

The relief valve must drip when the differential gauge is 2 psid or above.

Restore System
Step #1—Close all TCs
Step #2—Remove hoses
Step #3—Open all valves on test kit to drain water
Step #4—Restore #2 shutoff valve to pretest state

The tester shall provide copies of the test results to the owner and other appropriate parties as required.

The tester shall maintain a copy for his/her records.

Three-Valve Test on an ASSE 1015

Double-Check Backflow Prevention Assemblies (DC) and Double Check Fire Protection Backflow Prevention Assemblies (DCF)

Administrative Issues
A. Initial arrangements shall be made with the responsible party to schedule the test.
B. Arrangements/notifications shall be made where a continuous water supply is necessary or where testing creates a special hazard, inconvenience, or risk in buildings or for piping systems. Examples of such buildings or systems include, but are not limited by enumeration, to hospitals, manufacturing plants, fire protection systems, alarm companies/fire departments, insurance carriers, and others where special notification is necessary.
C. The proper information, reporting forms, and equipment shall be gathered to properly conduct the test.

Site Issues
A. Safety Inspection
The initial field evaluation shall be made for safety hazards in accordance with applicable federal, state, and local safety regulations and statutes. Some of the issues to be examined are, but are not limited by enumeration, to confined spaces, ventilation, access, oxygen content, chemical, electrical, or flammable hazards, hazards related to elevation of devices, hazards to the tester, and other persons.
B. Inspection of the Installation
The proper application of the assembly shall be confirmed for code compliance with respect to the degree of hazard, markings, prohibited locations (i.e., where subjected to flooding or freezing) and special installation requirements. The assembly orientation and direction of flow shall be confirmed as proper. The assembly shall be checked for alterations or special needs such as, but not limited by enumeration, to the evidence of illegal bypasses. The general appearance of the device shall be checked for the condition of the shutoff valves, test cocks, etc.

Field Testing Requirement

Flush Test Cocks (TC)
Step #1—Install test adapters (if applicable)
Step #2—Open TC #4—let trickle
Step #3—Open TC #1—close
Step #4—Open TC #2—close
Step #5—Open TC #3—close
Step #6—Close TC #4

Attach Test Kit
Step #1—Close high and low valves and open bypass valve on test kit
Step #2—Attach high hose to TC #2
Step #3—Attach low hose to TC #3
Step #4—Open TC #2
Step #5—Open high valve—bleed air—close
Step #6—Open TC #3
Step #7—Open low valve—bleed air—close
Step #8—Attach bypass hose to TC #4
Step #9—Open low valve
Step #10—Loosen bypass hose at TC #4—bleed air—tighten

Step #11—Close low valve
Step #12—Open high valve

Test #1—Tightness of #2 Shutoff Valve

Step #1—Close #2 shutoff valve
Step #2—Open TC #4
Step #3—Close TC #2 (pause to allow gauge to readjust)
Step #4—Read differential gauge

- Test Results -

If differential gauge reading remains steady, record #2 shutoff as tight.

Test #2—Tightness of #1 Check Valve

Step #1—Close TC #4
Step #2—Close high valve
Step #3—Remove bypass hose from TC #4
Step #4—Open TC #2
Step #5—Open low valve to cause differential reading to rise—close

- Test Results -

Record gauge value.

If differential reading holds steady at 1 psid or higher, record #1 check valve as tight.

Test #3—Tightness of #2 Check Valve

Step #1—Close TC #2 and TC #3
Step #2—Remove high and low hoses
Step #3—Attach high hose to TC #3
Step #4—Attach low hose to TC #4
Step #5—Open TC #3
Step #6—Open high valve—bleed air—close
Step #7—Open TC #4
Step #8—Open low valve—bleed air—close

- Test Results -

Record gauge value.

If differential reading holds steady at 1 psid or higher, record #2 check as tight.

Restore System

Step #1—Close all TCs
Step #2—Remove hoses
Step #3—Open all valves on test kit to drain water
Step #4—Restore #2 shutoff valve to pretest state

The tester shall provide copies of the test results to the owner and other appropriate parties as required.

The tester shall maintain a copy for his/her records.

Five-Valve Test on an ASSE 1015

Double-Check Backflow Prevention Assemblies (DC) and Double Check Fire Protection Backflow Prevention Assemblies (DCF)

Administrative Issues
A. Initial arrangements shall be made with the responsible party to schedule the test.
B. Arrangements/notifications shall be made where a continuous water supply is necessary or where testing creates a special hazard, inconvenience, or risk in buildings or for piping systems. Examples of such buildings or systems include, but are not limited by enumeration, to hospitals, manufacturing plants, fire protection systems, alarm companies/fire departments, insurance carriers, and others where special notification is necessary.
C. The proper information, reporting forms and equipment shall be gathered to properly conduct the test.

Site Issues
A. Safety Inspection
   The initial field evaluation shall be made for safety hazards in accordance with applicable federal, state, and local safety regulations and statutes. Some of the issues to be examined are, but are not limited by enumeration, to confined spaces, ventilation, access, oxygen content, chemical, electrical, or flammable hazards, hazards related to elevation of devices, hazards to the test, and other persons.
B. Inspection of the Installation
   The proper application of the assembly shall be confirmed for code compliance with respect to the degree of hazard, markings, prohibited locations (i.e., where subjected to flooding or freezing), and special installation requirements. The assembly orientation and direction of flow shall be confirmed as proper. The assembly shall be checked for alterations or special needs such as, but not limited by enumeration, to the evidence of illegal bypasses. The general appearance of the device shall be checked for the condition of shutoff valves, test cocks, etc.

Field Testing Requirements

Flush Test Cocks (TC)
Step #1—Install test adapters (if applicable)
Step #2—Open TC #4—let trickle
Step #3—Open TC #1—close
Step #4—Open TC #2—close
Step #5—Open TC #3—close
Step #6—Close TC #4

Attach Test Kit
Step #1—Close high, low and bypass valves and high and low bleed valves on test kit
Step #2—Attach high hose to TC #2
Step #3—Attach low hose to TC #3
Step #4—Open TC #2
Step #5—Open high bleed valve—bleed air—close
Step #6—Open TC #3
Step #7—Open low bleed valve—bleed air—close
Step #8—Attach bypass hose to TC #4
Step #9—Open high valve
Step #10—Open bypass valve
Step #11—Loosen bypass at TC #4—bleed air—tighten
Step #12—Slowly open low bleed valve to cause differential reading to rise—change

Test #1—Tightness of #2 Shutoff Valve

    Step #1—Close #2 shutoff valve
    Step #2—Open TC #4
    Step #3—Close TC #2 (pause to allow gauge to readjust)
    Step #4—Read differential gauge

- Test Results -

If differential gauge reading remains steady, record #2 shutoff as tight.

Test #2—Tightness of #1 Check Valve

    Step #1—Close TC #4
    Step #2—Close high valve
    Step #3—Remove bypass hose from TC #4
    Step #4—Open TC #2
    Step #5—Slowly open low bleed valve to cause differential reading to rise—change

- Test Results -

Record gauge value.

If differential reading holds steady at 1 psid or higher, record #1 check valve as tight.

Test #3—Tightness of #2 Check Valve

    Step #1—Close TC #2 and TC #3
    Step #2—Remove high and low hoses
    Step #3—Attach high hose to TC #3
    Step #4—Attach low hose to TC #4
    Step #5—Open TC #3
    Step #6—Open high bleed valve—bleed air—close
    Step #7—Open TC #4
    Step #8—Open low bleed valve—bleed air—change

- Test Results -

Record gauge value.

If differential reading holds steady at 1 psid or higher, record #2 check valve as tight.

Restore System

    Step #1—Close all TCs
    Step #2—Remove hoses
    Step #3—Open all valves on test kit to drain water
    Step #4—Restore #2 shutoff valve to pretest state

The tester shall provide copies of the test results to the owner and other appropriate parties as required.

The tester shall maintain a copy for his/her records.

Three-Valve Test on an ASSE 1020
Pressure Vacuum Breaker Assembly (PVB)

Administrative Issues
A. Initial arrangements shall be made with the responsible party to schedule the test.
B. Arrangements/notifications shall be made where a continuous water supply is necessary or where testing creates a special hazard, inconvenience, or risk in buildings or for piping systems. Examples of such buildings or systems include, but are not limited by enumeration, to hospitals, manufacturing plants, fire protection systems, alarm companies/fire departments, insurance carriers, and others where special notification is necessary.
C. The proper information, reporting forms, and equipment shall be gathered to properly conduct the test.

Site Issues
A. Safety Inspection

The initial field evaluation shall be made for safety hazards in accordance with applicable federal, state, and local safety regulations and statutes. Some of the issues to be examined are, but are not limited by enumeration, to confined spaces, ventilation, access, oxygen content, chemical, electrical, or flammable hazards, hazards related to elevation of devices, hazards to the tester, and other persons.

B. Inspection of the Installation

The proper application of the assembly shall be confirmed for code compliance with respect to the degree of hazard, markings, prohibited locations (i.e. where subjected to flooding or freezing), and special installation requirements. The assembly orientation and direction of flow shall be confirmed as proper. The assembly shall be checked for alterations or special needs such as, but not limited by enumeration, to the evidence of illegal bypasses and the adequacy of drainage systems from the assembly. The general appearance of the device shall be checked for evidence of excessive discharge, condition of shutoff valves, test cocks, and adequacy of drainage, should leakage occur.

Field Testing Requirements

Flush Test Cocks (TC)
Step #1—Install test adapters (if applicable)
Step #2—Open TC #1—close
Step #3—Open TC #2—change

Attach Test Kit
Step #1—Close high and low valves and open bypass valve on test kit
Step #2—Attach high hose to TC #1
Step #3—Attach low hose to TC #2
Step #4—Open TC #1
Step #5—Open high valve—bleed air—close
Step #6—Open TC #2
Step #7—Open low valve—bleed air—change

Test #1—Tightness of #2 Shutoff Valve
Step #1—Close #2 shutoff valve
Step #2—Close bypass valve
Step #3—Open high valve
Step #4—Open low valve (differential will read zero)

Step #5—Close high valve
Step #6—Close low valve
Step #7—Close #1 shutoff valve

- Test Results -

If differential gauge reading does not rise above zero (0), record #2 shutoff valve as tight.

Test #2—Tightness of the Check Valve

Step #1—Open #1 shutoff valve
Step #2—Open bypass valve
Step #3—Open low valve—bleed—close low valve

- Test Results -

If there is a differential reading of 1 psid or higher after reading stabilizes, record check valve as tight.

Test #3—Air Inlet Opening

Step #1—Close TC #1 and TC #2
Step #2—Remove hoses from TC #1 and TC #2
Step #3—Attach high hose to TC #2
Step #4—Remove air inlet canopy/hood
Step #5—Open TC #2
Step #6—Open high valve—bleed air—close
Step #7—Close #1 shutoff
Step #8—Center differential gauge at TC #2
Step #9—Open high valve and record differential gauge reading when the air inlet canopy/ hood opens

- Test Results -

If air inlet is visibly open when differential reading is 1 psid or higher, record valve as passed.

Replace air inlet canopy/hood

Restore System

Step #1—Close TC #2
Step #2—Remove high hose
Step #3—Open all valves on test kit to drain water
Step #4—Restore to pretest state

The tester shall provide copies of the test results to the owner and other appropriate parties as required.

The tester shall maintain a copy for his/her records.

Five-Valve Test on an ASSE 1020
Pressure Vacuum Breaker Assembly (PVB)

Administrative Issues
A. Initial arrangements shall be made with the responsible party to schedule the test.
B. Arrangements/notifications shall be made where a continuous water supply is necessary or where testing creates a special hazard, inconvenience, or risk in buildings or for piping systems. Examples of such buildings or systems include, but are not limited by enumeration, to hospitals, manufacturing plants, fire protection systems, alarm companies/fire departments, insurance carriers, and others where special notification is necessary.
C. The proper information, reporting forms, and equipment shall be gathered to properly conduct the test.

Site Issues
A. Safety Inspection
   The initial field evaluation shall be made for safety hazards in accordance with applicable federal, state, and local safety regulations and statutes. Some of the issues to be examined are, but are not limited by enumeration, to confined spaces, ventilation, access, oxygen content, chemical, electrical, or flammable hazards, hazards related to elevation of devices, hazards to the tester, and other persons.
B. Inspection of the Installation
   The proper application of the assembly shall be confirmed for code compliance with respect to the degree of hazard, markings, prohibited locations (i.e., where subjected to flooding or freezing), and special installation requirements. The assembly orientation and direction of flow shall be confirmed as proper. The assembly shall be checked for alterations or special needs such as, but not limited by enumeration, to the evidence of illegal bypasses and the adequacy of drainage systems from the assembly. The general appearance of the device shall be checked for evidence of excessive discharge, condition of shutoff valves, test cocks, and adequacy of drainage, should leakage occur.

Field Testing Requirements

Flush Test Cocks (TC)
   Step #1—Install test adapters (if applicable)
   Step #2—Open TC #1 -close
   Step #3—Open TC #2—change

Attach Test Kit
   Step #1—Close high, low and bypass valves and high and low bleed valves on test kit
   Step #2—Attach high to TC #1
   Step #3—Attach low hose to TC #2
   Step #4—Open TC #1
   Step #5—Open high bleed valve—bleed air—close
   Step #6—Open TC #2
   Step #7—Open low bleed valve—bleed air—change

Test #1—Tightness of #2 Shutoff Valve
   Step #1—Close #2 shutoff valve
   Step #2—Open high valve
   Step #3—Open low valve (differential will read zero)
   Step #4—Close high valve

Step #5—Close low valve
Step #6—Close #1 shutoff valve

- Test Results -

If differential gauge reading does not rise above zero (0), record #2 shutoff as tight.

Test #2—Tightness of Check Valve
Step #1—Open #1 shutoff valve
Step #2—Open low bleed valve—close low bleed valve

- Test Results -

If differential gauge reading holds steady at 1 psid or higher, record check valve as tight.

Test #3—Air Inlet Opening
Step #1—Close TC #1 and TC #2
Step #2—Remove hoses from TC #1 and TC #2
Step #3—Attach high hose to TC #2
Step #4—Remove air inlet canopy/hood
Step #5—Open TC #2
Step #6—Open high bleed valve—bleed air—close
Step #7—Close #1 shutoff valve
Step #8—Center differential gauge at TC #2
Step #9—Open high bleed valve and record the differential gauge reading when the air inlet canopy/hood opens

- Test Results -

If air inlet is visibly open, when differential reading is 1 psid or higher, record valve as passed.

Replace air inlet canopy/hood

Restore System
Step #1—Close TC #2
Step #2—Remove high hose
Step #3—Open all valves on test kit to drain water
Step #4—Restore to pretest status

The tester shall provide copies of the test results to the owner and other appropriate parties as required.

The tester shall maintain a copy for his/her records.

Three-Valve Test on an ASSE 1047

Reduced-Pressure Detector Fire Protection Backflow Prevention Assemblies (RPDA–RPDF)

Administrative Issues
A. Initial arrangements shall be made with the responsible party to schedule the test.
B. Arrangements/notifications shall be made where a continuous water supply is necessary or where testing creates a special hazard, inconvenience, or risk in buildings or for piping systems. Examples of such buildings or systems include, but are not limited by enumeration, to hospitals, manufacturing plants, fire protection systems, alarm companies/fire departments, insurance carriers, and others where special notification is necessary.
C. The proper information, reporting forms, and equipment shall be gathered to properly conduct the test.

Site Issues
A. Safety Inspection
   The initial field evaluation shall be made for safety hazards in accordance with applicable federal, state and local safety regulations and statutes. Some of the issues to be examined are, but are not limited by enumeration, to confined spaces, ventilation, access, oxygen content, chemical, electrical, or flammable hazards, hazards related to elevation of devices, hazards to the tester, and other persons.
B. Inspection of the Installation
   The proper application of the assembly shall be confirmed for code compliance with respect to the degree of hazard, markings, prohibited locations (i.e., where subjected to flooding or freezing), and special installation requirements. The assembly orientation and direction of flow shall be confirmed as proper. The assembly shall be checked for alterations or special needs such as, but not limited by enumeration, to the adequacy of the air gap, the evidence of illegal bypasses and the adequacy of drainage systems from the assembly. The general appearance of the device shall be checked for evidence of excessive discharge, condition of shutoff valves test cocks, relief valve, air gap, and adequacy of drainage should leakage occur.

Field Testing Requirements

Mainline Valve

Flush Test Cocks (TC)
Step #1—Install test adapters (if applicable)
Step #2—Open TC #4—let trickle
Step #3—Open TC #1—close
Step #4—Open TC #2—close
Step #5—Open TC #3—close
Step #6—Close TC #4

Attach Test Kit
Step #1—Close high, low valves and open bypass valve on test kit
Step #2—Attach high hose to TC #2
Step #3—Attach low hose to TC #3
Step #4—Open TC #2
Step #5—Open high valve—bleed air—close
Step #6—Open TC #3
Step #7—Open low valve—bleed air—close

Step #8—Attach bypass hose to TC #4
Step #9—Open high valve
Step #10—Loosen bypass hose at TC #4- bleed air- tighten

Test #1—Tightness of the #2 Shutoff Valve

Step #1—Close #2 shutoff valve in mainline
Step #2—Close #2 shutoff in bypass line
Step #3—Open TC #4
Step #4—Close TC #2
Step #5—Read differential gauge

- Test Results -

If the differential gauge reading remains steady, record #2 shutoff as tight.

Test #2—#2 Check w/Back Pressure Test

If in Test #1, the differential gauge remains steady, record #2 check valve as tight.

Test #3—#1 Check Differential

Step #1—Close TC #4
Step #2—Close high valve
Step #3—Remove bypass from TC #4
Step #4—Open TC #2
Step #5—Open low valve to cause differential gauge reading to increase
Step #6—Close low valve

- Test Results -

Record the pressure differential across #1 check valve.

If the differential reading is 5 psid or above, record #1 check valve as tight.

Test #4—Relief Valve Opening

Step #1—Close bypass valve
Step #2—Open high valve
Step #3—Slowly open low valve

- Test Results -

Record relief valve psid opening point. The relief valve must drip when the differential gauge is 2 psid or above.

Test #5A—Bypass Assembly (1013)

Remove Test Kit From Mainline Valve

Attach Test Kit to bypass Assembly (1047)

Step #1—Close high, low valves and open bypass valve on test kit
Step #2—Attach high hose to TC #2
Step #3—Attach low hose to TC #3
Step #4—Open TC #2
Step #5—Open high valve—bleed air—close
Step #6—Open TC #3
Step #7—Open low valve—bleed air—close
Step #8—Attach bypass hose to TC #4
Step #9—Open high valve
Step #10—Loosen bypass hose at TC #4- bleed air- tighten

Repeat Test #1—Tightness of #2 shutoff Valve

Repeat Test #2—Check w/Back Pressure Test

Repeat Test #3—#1 Check Differential

Repeat Test #4—Relief Valve Opening

Or

Test #5B—Bypass Check Assembly (1047)
(Second Check Only)
>Step #1—Close high valve—open bypass valve
>Step #2—Attach high hose to TC #2 (mainline assembly)
>Step #3—Attach low hose to TC #1 on bypass check
>Step #4—Open TC #2 on mainline assembly
>Step #5—Open high valve—bleed air—close
>Step #6—Open TC #1 on bypass check assembly
>Step #7—Open low valve—bleed air—close
>Step #8—Attach bypass hose to TC #2 on bypass check assembly
>Step #9—Open high valve
>Step #10—Loosen bypass hose at TC #2—bleed air—tighten
>Step #11—Open TC #2 on bypass check assembly

- Test Results -

If differential gauge reading remains steady, record bypass check as tight.

Test #6—Bypass Water Meter
>Step #1—Open #2 shutoff valve on bypass line
>Step #2—Open TC #4 on mainline assembly

- Test Results -

Verify that the water meter indicates flow.

Restore System
>Step #1—Close all TCs
>Step #2—Remove hoses
>Step #3—Open all valves on test kit to drain water
>Step #4—Restore mainline and bypass #2 shutoff valve to pretest state

The tester shall provide copies of the test results to the owner and other appropriate parties as required.

The tester shall maintain a copy for his/her records.

Three-Valve Test on an ASSE 1048

Double Check Detector Fire Protection Backflow Prevention Assemblies (DCDA–DCDF)

Administrative Issues

A. Initial arrangements shall be made with the responsible party to schedule the test.
B. Arrangements/notifications shall be made where a continuous water supply is necessary or where testing creates a special hazard, inconvenience, or risk in buildings or for piping systems. Examples of such buildings or systems include, but are not limited by enumeration, to hospitals, manufacturing plants, fire protection systems, alarm companies/fire departments, insurance carriers, and others where special notification is necessary.
C. The proper information, reporting forms, and equipment shall be gathered to properly conduct the test.

Site Issues

A. Safety Inspection
   The initial field evaluation shall be made for safety hazards in accordance with applicable federal, state, and local safety regulations and statutes. Some of the issues to be examined are, but are not limited by enumeration, to confined spaces, ventilation, access, oxygen content, chemical, electrical, or flammable hazards, hazards related to elevation of devices, hazards to the tester, and other persons.
B. Inspection of the Installation
   The proper application of the assembly shall be confirmed for code compliance with respect to the degree of hazard, markings, prohibited locations (i.e., where subjected to flooding or freezing), and special installation requirements. The assembly orientation and direction of flow shall be confirmed as proper. The assembly shall be checked for alterations or special needs such as, but not limited by enumeration, to the evidence of illegal bypasses. The general appearance of the device shall be checked for the condition of shutoff valves, test cocks, etc.

Field Testing Requirements

Mainline Valve

Flush Test Cocks (TC)

Step #1—Install test adapters (if applicable)
Step #2—Open TC #4—let trickle
Step #3—Open TC #1—close
Step #4—Open TC #2—close
Step #5—Open TC #3—close
Step #6—Close TC #4

Attach Test Kits Mainline Assembly

Step #1—Close high and low valves—open bypass valve on test kit
Step #2—Attach high hose to TC #2
Step #3—Attach low hose to TC #3
Step #4—Open TC #2
Step #5—Open high valve—bleed air—close
Step #6—Open TC #3
Step #7—Open low valve—bleed air—close
Step #8—Attach bypass hose to TC #4
Step #9—Open low valve

Step #10—Loosen bypass hose at TC #4- bleed air- tighten
Step #11—Close low valve
Step #12—Open high valve

Test #1—Tightness of #2 Shutoff Valve
Step #1—Close #2 shutoff on mainline
Step #2—Close #2 shutoff on bypass line
Step #3—Open TC #4
Step #4—Close TC #2
Step #5—Read differential gauge

- Test Results -

If differential gauge reading holds steady, record #2 shutoff valve as tight.

Test #2—Tightness of #1 Check Valve
Step #1—Close TC #4
Step #2—Close high valve
Step #3—Remove bypass hose from TC #4
Step #4—Open TC #2
Step #5—Open low valve to cause differential reading to rise—close

- Test Results -

If differential reading holds steady at 1.0 psid or higher for the DCDA or 0.5 psid or higher for DCDF, record #1 check valve as tight.

Test #3—Tightness of #2 Check Valve
Step #1—Close TC #2 and TC #3
Step #2—Remove high and low hoses
Step #3—Attach high hose to TC #3
Step #4—Attach low hose to TC #4
Step #5—Open TC #3
Step #6—Open high valve—bleed air—close
Step #7—Open TC #4
Step #8—Open low valve—bleed air—change

- Test Results -

If differential reading holds steady at 1.0 psid or higher, record #2 check valve as tight.

Step #4A -Bypass Assembly (1015)

Remove Test Kit from Mainline Valve

Attach Test Kit to Bypass Assembly (1048)
Step #1—Close high and low valves and open bypass on test kit
Step #2—Attach high hose to TC #2
Step #3—Attach low hose to TC #3
Step #4—Open TC #2
Step #5—Open high valve—bleed air—close
Step #6—Open TC #3
Step #7—Open low valve—bleed air—close
Step #8—Attach bypass hose to TC #4
Step #9—Open high valve
Step #10—Loosen bypass hose at TC #4- bleed air—change

Repeat Test #1—Tightness of #2 Shutoff Valve

Repeat Test #2—Tightness of #1 Check Valve

Repeat Test #3—Tightness of #2 Check Valve

or

Test #4B—Bypass Check Assembly (1048)

(Second Check Only)
- Step #1—Attach high hose to TC #1 on bypass check and low hose to TC #2 on bypass check
- Step #2—Open TC #1 and TC #2 on bypass check
- Step #3—Bleed high valve—close when air is released
- Step #4—Bleed low valve—close when air is released

- Test Result -

If the differential reading holds steady at 1.0 psid or higher, record the bypass check as tight.

Test #5—Bypass Water Meter

- Step #1—Open #2 shutoff valve on bypass line
- Step #2—Open TC #4 on mainline assembly

- Test Results -

Verify that the water meter indicates flow.

Restore System

- Step #1—Close all TCs
- Step #2—Remove hoses
- Step #3—Open all valves on test kit to drain water
- Step #4—Restore mainline and bypass #2 shutoff valve to pretest state

The tester shall provide copies of the test results to the owner and other appropriate parties as required.

The tester shall maintain a copy for his/her records.

Five-Valve Test on an ASSE 1048

Double Check Detector Fire Protection Backflow Prevention Assemblies (DCDA–DCDF)

Administrative Issues

A. Initial arrangements shall be made with the responsible party to schedule the test.
B. Arrangements/notifications shall be made where a continuous water supply is necessary or where testing creates a special hazard, inconvenience, or risk in buildings or for piping systems. Examples of such buildings or systems include, but are not limited by enumeration, to hospitals, manufacturing plants, fire protection systems, alarm companies/fire departments, insurance carriers, and others where special notification is necessary.
C. The proper information, reporting forms, and equipment shall be gathered to properly conduct the test.

Site Issues

A. Safety Inspection

The initial field evaluation shall be made for safety hazards in accordance with applicable federal, state, and local safety regulations and statutes. Some of the issues to be examined are, but are not limited by enumeration, to confined spaces, ventilation, access, oxygen content, chemical, electrical, or flammable hazards, hazards related to elevation of devices, hazards to the tester, and other persons.

B. Inspection of the Installation

The proper application of the assembly shall be confirmed for code compliance with respect to the degree of hazard, markings, prohibited locations (i.e., where subjected to flooding or freezing), and special installation requirements. The assembly orientation and direction of flow shall be confirmed as proper. The assembly shall be checked for alterations or special needs such as, but not limited by enumeration, to the evidence of illegal bypasses. The general appearance of the device shall be checked for the condition of shutoff valves, test cocks, etc.

Field Testing Requirements

Mainline Valve

Flush Test Cocks (TC)

Step #1—Install test adapters (if applicable)
Step #2—Open TC #4—let trickle
Step #3—Open TC #1—close
Step #4—Open TC #2—close
Step #5—Open TC #3—close
Step #6—Close TC #4

Attach Test Kit

Step #1—Close High, low and bypass valves and high and low bleed valves on test kit
Step #2—Attach high hose to TC #2
Step #3—Attach low hose to TC #3
Step #4—Open TC #2
Step #5—Open high bleed valve—bleed air—close
Step #6—Open TC #3
Step #7—Open low bleed valve—bleed air—close
Step #8—Attach bypass hose to TC #4
Step #9—Open high valve
Step #10—Open bypass valve
Step #11—Loosen bypass at TC #4—bleed air—tighten
Step #12—Slowly open low bleed valve to cause differential reading to rise—change

Test #1—Tightness of #2 Shutoff

Step #1—Close #2 shutoff valve on mainline
Step #2—Close #2 shutoff on bypass line
Step #3—Open TC #4
Step #4—Close TC #2
Step #5—Read differential gauge

- Test Results -

If differential gauge reading holds steady, record #2 shutoff as tight.

Test #2—Tightness of #1 Check Valve
    Step #1—Close TC #4
    Step #2—Close high valve
    Step #3—Remove bypass hose from TC #4
    Step #4—Open TC #2
    Step #5—Slowly open low bleed valve to cause differential reading to rise—change

- Test Results -

If differential reading holds steady at 1.0 psid or higher for the DCDA or 0.5 psid or higher for DCDF, record #1 check valve as tight.

Test #3—Tightness of #2 Check Valve
    Step #1—Close TC #2 and TC #3
    Step #2—Remove high and low hoses
    Step #3—Attach high hose to TC #3
    Step #4—Attach low hose to TC #4
    Step #5—Open TC #3 and TC #4
    Step #6—Open high bleed valve—bleed air—close
    Step #7—Open low bleed valve—bleed air—change

- Test Results -

If differential reading holds steady at 1.0 psid or higher, record #2 check valve as tight.

Test 4A—Bypass Assembly(1015)

Remove Test Kit from Mainline Valve

Attach Test Kit to bypass Assembly
    Step #1—Close High, low and bypass valves and high and low bleed valves on test kit
    Step #2—Attach high hose to TC #2
    Step #3—Attach low hose to TC #3
    Step #4—Open TC #2
    Step #5—Open high bleed valve—bleed air—close
    Step #6—Open TC #3
    Step #7—Open low bleed valve—bleed air—close
    Step #8—Attach bypass hose to TC #4
    Step #9—Open high valve
    Step #10—Open bypass valve
    Step #11—Loosen bypass at TC #4—bleed air—close
    Step #12—Slow open low bleed valve to cause differential reading to rise—change

Repeat Test #1—Tightness of #2 shutoff Valve

Repeat Test #2—Tightness of #1 Check Valve

Repeat Test #3—Tightness of #2 Check Valve

Or

Test #4B—Bypass Check Assembly (1048)

(Second Check Only)
    Step #1—Attach high hose to TC #1 on bypass check
    Step #2—Attach low hose to TC #2 on bypass check
    Step #3—Close high and low valves and high and low bleed valves
    Step #4—Open TC #1 and TC #2 on bypass check
    Step #5—Open high bleed to remove air—close
    Step #6—Open low bleed to remove air—close

- Test Results -

If the differential reading holds steady at 1.0 psid or higher, record the bypass check as tight.

Test #5—Bypass Water Meter
Step #1—Open #2 shutoff on bypass line
Step #2—Open TC #4 on mainline assembly

- Test Results -

Verify that the water meter indicates flow.

Restore System
Step #1—Close all TCs
Step #2—Remove hoses
Step #3—Open all valves on test kit to drain water
Step #4—Restore mainline and bypass #2 shutoff valve to pretest state

The tester shall provide copies of the test results to the owner and other appropriate parties as required.

The tester shall maintain a copy for his/her records.

Three-Valve Test on an ASSE 1056
Spill-Resistant Vacuum Breaker (SRVB)

Administrative Issues
A. Initial arrangements shall be made with the responsible party to schedule the test.
B. Arrangements/notifications shall be made where a continuous water supply is necessary or where testing creates a special hazard, inconvenience, or risk in buildings or for piping systems. Examples of such buildings or systems include, but are not limited by enumeration, to hospitals, manufacturing plants, fire protection systems, alarm companies/fire departments, insurance carriers, and others where special notification is necessary.
C. The proper information, reporting forms, and equipment shall be gathered to properly conduct the test.

Site Issues
A. Safety Inspection

The initial field evaluation shall be made for safety hazards in accordance with applicable federal, state, and local safety regulations and statutes. Some of the issues to be examined are, but are not limited by enumeration, to confined spaces, ventilation, access, oxygen content, chemical, electrical, or flammable hazards, hazards related to elevation of devices, hazards to the tester, and other persons.

B. Inspection of the Installation

The proper application of the assembly shall be confirmed for code compliance with respect to the degree of hazard, markings, prohibited locations (i.e., where subjected to flooding or freezing), and special installation requirements. The assembly orientation and direction of flow shall be confirmed as proper. The assembly shall be checked for alterations or special needs such as, but not limited by enumeration, to the evidence of illegal bypasses. The general appearance of the device shall be checked for evidence of excessive discharge, condition of shutoff valves, test cocks, and adequacy of drainage, should leakage occur.

Field Testing Requirements

Flush Test Cocks
    Step #1—Bleed test cock

Attach Test Kit
    Step #1—Attach test kit
    Step #2—Test kit shall be centered at TC #1 for all tests
    Step #3—Close high and low valves
    Step #4—Attach high to TC #1
    Step #5—Open TC #1
    Step #6—Bleed air by opening High bypass valve—change

Test #1—Tightness of Check Valve
    Step #1—Close #2 shutoff
    Step #2—Close #1 shutoff
    Step #3—Open bleed screw

- Test Results -

If the differential reading is 1 psid or higher when the discharge from the bleed screw stops, record as passed.

Test #2—Air Inlet Opening
    Step #1—Remove air inlet canopy
    Step #2—Open high valve to reduce differential reading to 1 psid

- Test Results -

If the air inlet is visually open when the differential gauge reads 1 psid, record as open.

Restore System
    Step #1—Close all TCs
    Step #2—Remove hoses
    Step #3—Open all valves on test kit to drain water
    Step #4—Restore to pretest state

The tester shall provide copies of the test results to the owner and other appropriate parties as required.

The tester shall maintain a copy for his/her records.

Five-Valve Test on an ASSE 1056
    Spill-Resistant Vacuum Breaker (SRVB)

Administrative Issues
A. Initial arrangements shall be made with the responsible party to schedule the test.
B. Arrangements/notifications shall be made where a continuous water supply is necessary or where testing creates a special hazard, inconvenience, or risk in buildings or for piping systems. Examples of such buildings or systems include, but are not limited by enumeration, to hospitals, manufacturing plants, fire protection systems, alarm companies/fire departments, insurance carriers, and others where special notification is necessary.
C. The proper information, reporting forms, and equipment shall be gathered to properly conduct the test.

Site Issues
A. Safety Inspection

The initial field evaluation shall be made for safety hazards in accordance with applicable federal, state, and local safety regulations and statutes. Some of the issues to be examined are, but are not limited by enumeration, to confined spaces, ventilation, access, oxygen content, chemical, electrical, or flammable hazards, hazards related to elevation of devices, hazards to the tester, and other persons.

B. Inspection of the Installation

The proper application of the assembly shall be confirmed for code compliance with respect to the degree of hazard, markings, prohibited locations (i.e., where subjected to flooding or freezing), and special installation requirements. The assembly orientation and direction of flow shall be confirmed as proper. The assembly shall be checked for alterations or special needs such as, but not limited by enumeration, to the evidence of illegal bypasses. The general appearance of the device shall be checked for evidence of excessive discharge, condition of shutoff valves, test cocks, and adequacy of drainage, should leakage occur.

Field Testing Requirements

Flush Test Cock
Step #1—Bleed test cock

Attach Test Kit
Step #1—Attach test kit
Step #2—Test kit shall be centered at TC #1 for all tests
Step #3—Close high and low valve and high and low bleed valves—open bypass
Step #4—Attach high to TC #1
Step #5—Open TC #1
Step #6—Bleed air by opening high bleed valve—change

Test #1—Tightness of Check Valve/Shutoff Valves
Step #1—Close #2 shutoff
Step #2—Close #1 shutoff
Step #3—Open bleed screw on SVB

- Test Results -
If the differential reading is 1 psid or higher when the discharge from the bleed screw stops, record as passed.

Test #2—Air Inlet Opening
Step #1—Remove air inlet canopy
Step #2—Open high valve to reduce differential reading to 1 psid

- Test Results -
If the air inlet is visibly open when the differential gauge reads 1 psid, record as open.

Restore System
Step #1—Close all TCs
Step #2—Remove hoses
Step #3—Open all valves on test kit to drain water
Step #4—Restore to pretest state

The tester shall provide copies of the test results to the owner and other appropriate parties as required.

The tester shall maintain a copy for his/her records.

One-Hose Test Procedure for an ASSE 1013

Reduced-Pressure Principle Backflow Preventers (RP) and Reduced-Pressure Principle Fire Protection Backflow Preventers (RPF)

Administrative Issues
A. Initial arrangements shall be made with the responsible party to schedule the test.
B. Arrangements/notifications shall be made where a continuous water supply is necessary or where testing creates a special hazard, inconvenience, or risk in buildings or for piping systems. Examples of such buildings or systems include, but are not limited by enumeration, to hospitals, manufacturing plants, fire protection systems, alarm companies/fire departments, insurance carriers, and others where special notification is necessary.
C. The proper information, reporting forms, and equipment shall be gathered to properly conduct the test.

Site Issues
A. Safety Inspection
   The initial field evaluation shall be made for safety hazards in accordance with applicable federal, state, and local safety regulations and statutes. Some of the issues to be examined are, but are not limited by enumeration, to confined spaces, ventilation, access, oxygen content, chemical, electrical, or flammable hazards, hazards related to elevation of devices, hazards to the tester, and other persons.
B. Inspection of the Installation
   The proper application of the assembly shall be confirmed for code compliance with respect to the degree of hazard, markings, prohibited locations (i.e., where subjected to flooding or freezing), and special installation requirements. The assembly orientation and direction of flow shall be confirmed as proper. The assembly shall be checked for alterations or special needs such as, but not limited by enumeration, to the adequacy of the air gap, the evidence of illegal bypasses, and the adequacy of drainage systems from the assembly. The general appearance of the device shall be checked for evidence of excessive discharge, condition of shut-off valves, test cocks, relief valve, air gap, and adequacy of drainage, should leakage occur.

Field Testing Requirements

Flush Test Cocks (TC)
Step #1—Install gauge adapters (if applicable)
Step #2—Open TC #4
Step #3—Open TC #2 and close
Step #4—Open TC #3 and close
Step #5—Close TC #4

Attach Test Kit
   Step #1—Close gauge manifold control valves and low-pressure bleed
   Step #2—Open gauge high-bleed valve
   Step #3—Open TC #2 with trickle flow and attach gauge high-pressure hose
   Step #4—Close gauge high-bleed valve and open TC #2 fully

Test #1—Check Valve #1 Differential
   Step #1—Close shutoff #2 then close shutoff #1
   Step #2—Slowly open TC #3
   Step #3—When water stops flowing from TC #3, record the gauge value for check valve #1

– Test Results –

Record the pressure differential across #1 check valve.

If the differential reading is 5 psid or above, record #1 check valve as tight.

Test #2—Relief Valve Opening
   Step #1—Slowly open gauge high-pressure bleed valve
   Step #2—Feel for relief valve discharge and record gauge value at opening
   Step #3—Continue to open high-bleed valve and verify that relief valve fully opens
   Step #4—Close TC #2 and TC #3 and remove high-pressure hose

– Test Results –

The relief valve must drip when the differential gauge is 2 psid.

Test #3—Check Valve #2 Differential
   Step #1—Reestablish pressure to assembly
      a. Open shutoff #1 or
      b. Attach bypass hose from TC #1 to TC #2 and open both test cocks
   Step #2—Open TC #3 with trickle flow and attach gauge high-pressure hose
   Step #3—Close gauge high-bleed valve and open TC #3 fully
   Step #4—Close shutoff #1 or close TC #2
   Step #5—Slowly open TC #4
   Step #6—When water stops flowing from TC #4, record the gauge value for check valve #2
   Step #7—Close TC #3 and TC #4. Remove high-pressure hose.
   Step #8—Restore assembly to pretest condition

– Test Results –

Record the pressure differential across #2 check valve.

If the differential reading is 1.0 psid or above, record #2 check valve as tight.

One-Hose Test Procedure for an ASSE 1015
   Double Check Backflow Prevention Assemblies (DC) and Double Check Fire Protection Backflow Prevention Assemblies (DCF)

Administrative Issues
A. Initial arrangements shall be made with the responsible party to schedule the test.
B. Arrangements/notifications shall be made where a continuous water supply is necessary or where testing creates a special hazard, inconvenience, or risk in buildings or for piping systems. Examples of such buildings or systems include, but are not limited by enumeration, to hospitals, manufacturing plants, fire protection systems, alarm

companies/fire departments, insurance carriers, and others where special notification is necessary.
C. The proper information, reporting forms, and equipment shall be gathered to properly conduct the test.

Site Issues

A. Safety Inspection

The initial field evaluation shall be made for safety hazards in accordance with applicable federal, state, and local safety regulations and statutes. Some of the issues to be examined are, but are not limited by enumeration, to confined spaces, ventilation, access, oxygen content, chemical, electrical, or flammable hazards, hazards related to elevation of devices, hazards to the tester, and other persons.

B. Inspection of the Installation

The proper application of the assembly shall be confirmed for code compliance with respect to the degree of hazard, markings, prohibited locations (i.e., where subjected to flooding or freezing), and special installation requirements. The assembly orientation and direction of flow shall be confirmed as proper. The assembly shall be checked for alterations or special needs such as, but not limited by enumeration, to the adequacy of the air gap, the evidence of illegal bypasses, and the adequacy of drainage systems from the assembly. The general appearance of the device shall be checked for evidence of excessive discharge, condition of shut-off valves, test cocks, relief valve, air gap, and adequacy of drainage, should leakage occur.

Field Testing Requirements

Flush Test Cocks (TC)

Step #1—Install gauge adapters (if applicable)
Step #2—Open TC #4
Step #3—Open TC #2 and close
Step #4—Open TC #3 and close
Step #5—Close TC #4

Attach Test Kit

Step #1—Close gauge manifold control valves and low-pressure bleed
Step #2—Open gauge high-bleed valve
Step #3—Open TC #2 with trickle flow and attach gauge high-pressure hose
Step #4—Close gauge high-bleed valve and open TC #2 fully

Test #1—Check Valve #1 Differential

Step #1—Close shutoff #2 then close shutoff #1
Step #2—Slowly open TC #3
Step #3—When water stops flowing from TC #3, record the gauge value for check valve #1
Step #4—Close TC #2 and TC #3 and remove high-pressure hose

- Test Results -

Record the pressure differential across #1 check valve.

If the differential reading is 1 psid or above, record #1 check valve as tight.

Test #2—Check Valve #2 Differential

Step #1—Reestablish pressure to assembly
    a. Open shutoff #1, or
    b. Attach bypass hose from TC #1 to TC #2 and open both test cocks
Step #2—Open TC #3 with trickle flow and attach gauge high-pressure hose

Step #3—Close gauge high-bleed valve and open TC #3 fully
Step #4—Close shutoff #1, or close TC #2
Step #5—Slowly open TC #4
Step #6—When water stops flowing from TC #4, record the gauge value for check valve #2
Step #7—Close TC #3 and TC #4. Remove high-pressure hose.
Step #8—Restore assembly to pretest condition

- Test Results -

Record the pressure differential across #2 check valve.

If the differential reading is 1 psid or above, record #2 check valve as tight.

One-Hose Test Procedure for an ASSE 1020
Pressure Vacuum Breaker Assembly (PVB)

Administrative Issues
A. Initial arrangements shall be made with the responsible party to schedule the test.
B. Arrangements/notifications shall be made where a continuous water supply is necessary or where testing creates a special hazard, inconvenience, or risk in buildings or for piping systems. Examples of such buildings or systems include, but are not limited by enumeration, to hospitals, manufacturing plants, fire protection systems, alarm companies/fire departments, insurance carriers, and others where special notification is necessary.
C. The proper information, reporting forms, and equipment shall be gathered to properly conduct the test.

Site Issues
A. Safety Inspection

The initial field evaluation shall be made for safety hazards in accordance with applicable federal, state, and local safety regulations and statutes. Some of the issues to be examined are, but are not limited by enumeration, to confined spaces, ventilation, access, oxygen content, chemical, electrical, or flammable hazards, hazards related to elevation of devices, hazards to the tester, and other persons.

B. Inspection of the Installation

The proper application of the assembly shall be confirmed for code compliance with respect to the degree of hazard, markings, prohibited locations (i.e., where subjected to flooding or freezing), and special installation requirements. The assembly orientation and direction of flow shall be confirmed as proper. The assembly shall be checked for alterations or special needs such as, but not limited by enumeration, to the evidence of illegal bypasses and the adequacy of drainage systems from the assembly. The general appearance of the device shall be checked for evidence of excessive discharge, condition of shutoff valves, test cocks, and adequacy of drainage, should leakage occur.

Field Testing Requirements

Flush Test Cocks (TC)
Step #1—Install test adapters (if applicable)
Step #2—Open TC #1—close
Step #3—Open TC #2—close
Step #4—Remove canopy from air-inlet valve

Step #5—Verify the orientation of each shutoff valve. #1 shutoff must be open and #2 shutoff may be closed at this time.

Attach Test Kit—Air Inlet Valve

Step #1—Close all the gauge valves

Step #2—Open gauge high-bleed valve

Step #3—Open TC #2 with trickle flow and attach the gauge high-pressure hose to TC #2

Step #4—Once water flows from the bleed/bypass hose, SLOWLY close gauge high-bleed valve

Step #5—Open TC #2 completely

Test #1—Air Inlet Valve Opening

Step #1—Verify that #2 shutoff is closed and then close #1 shutoff

Step #2—Establish gauge centerline with assembly air-inlet valve and maintain through Step 4

Step #3—Slightly open gauge high-bleed valve

Step #4—Test Results- Observe and record the value at the air-inlet valve opening (NOTE: this value must be 1.0 psid or greater.)

Step #5—Close TC #2 and remove gauge hose

Step #6—Open #1 shutoff

Step #7—Reinstall canopy

Attach Test Kit—Check Valve

Step #1—Close all the valves

Step #2—Open gauge high-bleed valve

Step #3—Open TC #1 with trickle flow and attach the high-pressure hose to TC #1

Step #4—Once water flows from the bleed/bypass hose, SLOWLY close gauge high-bleed valve

Step #5—Open TC #1 completely

Test #2—Tightness of Check Valve

Step #1—Close #1 shutoff valve

Step #2—Establish gauge centerline with assembly check valve and maintain through Step 4.

Step #3—Slowly open TC #2

Step #4—Test Results—Observe and record the value when water stops flowing from the test cock. (NOTE: This value must be 1.0 psid or greater.)

Step #5—Close TC #1 and TC #2

Step #6—Restore assembly to pretest condition.

INFORMATION PROVIED BY UNIVERSITY OF FLORIDA TRAINING, RESEARCH AND EDUCATION FOR ENVIRONMENTAL OCCUPATIONS (TREEO)

DCVA Field Test With Differential Gauge—Single Hose

Preparation
Notify customer
Inspect area
Flush test cocks
Install fittings
Inspect test kit—close all needle valves

CV 1
(Figure B-28)
Install compensating tee (bleed valve) on test cock 2
Install short clear tube on test cock 3
Install test gauge and end of low hose at same height water discharges from short clear tube
Attach high pressure hose to bleed valve on test cock 2
Open test cock 3 to fill clear tube
Close test cock 3
Open test cock 2 slowly
Open high bleed—bleed air from gauge
Close high bleed
Close outlet shutoff valve
Close inlet shutoff valve
Open test cock 3 (test cock 2 must be open)
NOTE: Gauge must read 1.0 psi or greater to pass
Record value of check valve #1

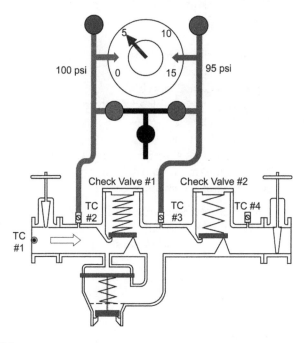

Figure 8-28 Test 1—CV#1

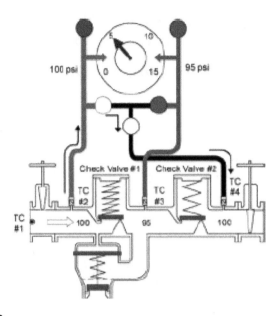

Figure 8-29 Test 3—CV#2

CV 2
(Figure B-29)
Close test cock 2 and test cock 3
Move short tube from test cock 3 to test cock 4
Remove high hose from test cock 2
Open inlet shutoff valve
Attach high hose to test cock 3 (use the compensating tee if inlet shutoff leaks)
Open test cock 4 to fill tube
Close test cock 4
Open test cock 3 slowly
Open high bleed—bleed air from gauge
Close high bleed
Close inlet shutoff valve
Open test cock 4 (test cock 3 must be open)
Note: gauge must read 1.0 psi or greater to pass
Record value of check valve #2

Final
Close test cocks—remove all equipment
Open inlet shutoff valve
Open outlet shutoff valve slowly
Open all needle valves on test kit

PVB Field Test with Differential Pressure Gauge

Preparation
Notify customer
Inspect area
Remove canopy
Flush test cocks
Install fittings
Inspect test equipment—close all needle valves

Air Inlet
Install test gauge and end of low pressure hose level with test cock 2
Attach high-pressure hose to test cock 2
Open test cock 2 slowly
Open high bleed—bleed air from gauge
Close high bleed
Close outlet shutoff valve
Close inlet shutoff valve
Finger in top—open high bleed (¼ turn max.)
Record when air inlet opens equal to or greater than 1.0 psi

> NOTE: If pressure will not drop on gauge, open test cock 1 slowly until pressure on gauge drops. (the inlet shutoff valve may be leaking.)

Check Valve
Close test cock 2
Remove high pressure hose from test cock 2
Open inlet shutoff valve
Attach compensating (bleed-off) tee to test cock 1
Attach high pressure hose to compensating (bleed-off) tee
Open test cock 1 slowly
Open high bleed—bleed air from gauge
Close high bleed
Close inlet shutoff valve
Open test cock 2
When water stops running from test cock 2
Record check valve equal to or greater than 1.0 psi

Final
Close test cocks—remove all equipment
Replace canopy
Open all needle valves on test kit
Open inlet shutoff valve
Open outlet shutoff valve slowly

Reduced-Pressure Principle Field Test with Five-Valve Test Kit

Preparation	Notify customer
	Inspect area
	Flush test cocks *(open 4, 3, 2, 1, then close 1, 2, 3, 4)*
	Install fittings
	Inspect test kit—close all needle valves
Observe	Attach high pressure hose to test cock 2
Check	Attach low pressure hose to test cock 3
Valve #1	Open test cock 3 slowly then open low bleed
	Open test cock 2 **slowly** then open high bleed
	Close bleeds—high first, low last close outlet shut-off valve
	Observe check valve #1—(record as closed tight or leaking)
Record	Open high by-pass control needle valve 1 full turn
Relief	Open low by-pass control *slightly*—no more than ¼ turn
Valve	Record RV opening > or = 2.0 psi
	Close low by-pass control needle valve
Observe	Bleed vent (by-pass) hose
	Check
	Attach vent hose to test cock 4 Valve #2.
	Close vent (by-pass) control leaks or open test cock 4
Closed tight	Reset gauge—(low bleed) to re-establish differential pressure across CV #1
	Open vent control one full turn
	Observe whether relief valve vent drips
	*(if the relief vent drips, reset gauge [low bleed], if relief valve drips a **second time**, then check valve 2 has failed and must be repaired)*
	(record as closed tight or leaking) stop test here if CV #2 is leaking.
Record check	Reset gauge—(low bleed) to relieve disk compression
Valve #1	Record check valve #1 differential (min. 5.0 psi and > RV opening)
Record outlet	Close test cock 2—**wait** and check gauge for leaks in outlet shutoff valve
Shutoff valve	(record as closed tight or leaking)

If the needle on the gauge holds steady there is no flow through the assembly, continue to next test.

Record	Close vent control
Check	Close test cocks 3 and 4
Valve #2	Remove vent hose from test cock 4
	Move low pressure hose to test cock 4
	Move high pressure hose to test cock 3
	Open test cock 4 slowly then open low bleed
	Open test cock 3 slowly then open high bleed
	Close high bleed first, close low bleed slowly
	Record CV 2 differential > or = 1.0 psi
Final	Close test cocks—remove all equipment
	Open all needle valves on test kit
	Open outlet shut-off valve slowly

RP Field Test with a Two-Valve Test Kit (Figure B-30)

Preparation	Notify customer
	Inspect area
	Flush test cocks *(open 4, 3, 2, 1, then close 1, 2, 3, 4)*
	Install fittings
	Inspect test kit—close all needle valves
Observe	Attach high pressure hose to test cock 2
Check	
	Attach low pressure hose to test cock 3
Valve #1	Open test cock 3 slowly then open low bleed needle valve
	Open test cock 2 **slowly** then open high bleed needle valve
	Close bleeds—high first, low last
	Close outlet shut-off valve
	Observe check valve #1—(<u>record</u> as closed tight or leaking)
Record	Attach a third hose (vent hose) to the high bleed needle valve
Relief	Open high bleed *slightly* to bleed air from vent hose
Valve	Connect other end of vent hose to low bleed needle valve
	Open high bleed valve 1 full turn
	Open low bleed slightly—no more than 1/4 turn
	<u>Record</u> RV opening > or = 2.0 psi
	Close low bleed needle valve—change high bleed needle valve
Observe	Disconnect vent hose from low bleed needle valve
Check	Bleed vent hose to remove air
Valve #2	Attach the vent hose to test cock 4 Leaks or
	Close high bleed needle valve closed tight
	Open test cock 4
	Reset gauge—(open then close low bleed)
	Open high bleed needle valve one full turn.
	Observe whether relief valve vent drips
	(if the relief vent drips, reset gauge [low bleed], if relief valve drips a second time, then check valve 2 has failed and must be repaired)
	(<u>record</u> as closed tight or leaking) stop test here if cv#2 is leaking.
Record check	Reset gauge—(low bleed)
Valve #1	<u>Record</u> CV 1 differential (5.0 psi min. and > RV opening)
Record outlet	Close test cock 2—wait and check gauge for leaks in outlet shutoff valve shutoff valve
	(<u>record</u> as closed tight or leaking)

If the needle on the gauge holds steady there is no flow through the assembly, continue to next test.

Record check	Close high bleed valve Valve #2
	Close test cocks 3 and 4
	Remove vent hose from test cock 4 Move low hose to test cock 4
	Move high hose to test cock 3
	Open test cock 4 slowly then open low bleed
	Open test cock 3 slowly then open high bleed
	Close high bleed first, close low bleed slowly
	<u>Record</u> check valve 2 differential > or = 1.0 psi
Final	Close test cocks—remove all equipment
	Open all needle valves on test kit
	Open outlet shut-off valve slowly

176 BACKFLOW PREVENTION AND CROSS-CONNECTION CONTROL

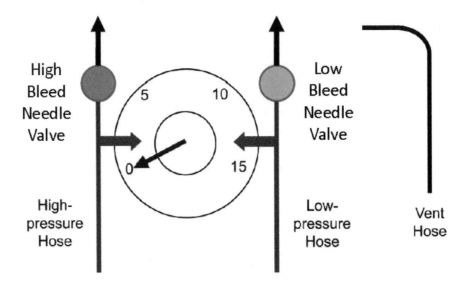

Figure 8-30 Two-valve differential pressure test kit

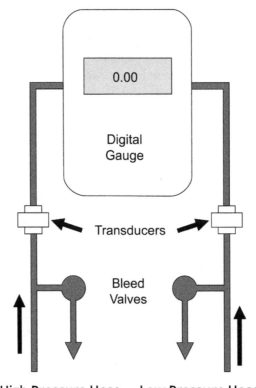

Figure 8-31 Two-valve test kit

Reduced-Pressure Principle Field Test with a Three-Valve Test Kit

Preparation	Notify customer
	Inspect area
	Flush test cocks *(open 4, 3, 2, 1, then close 1, 2, 3, 4)*
	Install fittings
	Inspect test kit—close all needle valves
Observe CV 1	Attach high hose to test cock 2
	Attach low hose to test cock 3
	Open test cock 3 **slowly** then open low control valve and vent control valve
	Open test cock 2 slowly then open high control valve
	Close high control first, close low control valve
	Close vent control valve
	Close outlet shut-off valve
	Observe CV 1—(record as closed tight or leaking)
Record	Open high control valve 1 full turn
Relief	Open low control *slightly*—no more than ¼ turn Valve Hand under vent
	Record RV opening > or = 2.0 psi. Close low control valve
Observe CV 2	Open high control valve. Open vent control valve slightly
	Leaks or
	Attach vent hose to test cock 4
Closed	Close vent control valve
Tight	Open test cock 4
	Reset gauge—(open then close low hose connection at test cock 3)
	Open vent control valve one full turn
	Observe whether relief valve vent drips
	(if the relief vent drips, reset gauge , if relief valve drips a **second time**, then check valve 2 has failed and must be repaired)
	(Record as closed tight or leaking) stop test here if cv#2 is leaking.
Record CV 1	Reset gauge—(open then close low hose connection at test cock 3)
	Record CV 1 differential (5.0 psi and > RV opening)
Record outlet	Close test cock 2 – **wait** and check gauge for leaks in outlet shut-off valve shutoff valve
	(record as closed tight or leaking)
	If the needle on the gauge holds steady there is no flow through the assembly, continue to next test.
Record CV 2	Close high control valve and vent control valve Close test cocks 3 and 4
	Remove vent hose from test cock 4. Move low hose to test cock 4.
	Move high hose to test cock 3
	Open test cock 4 slowly then open low control valve, and vent control valve
	Open test cock 3 slowly then open high control valve
	Close high control valve first; close low control valve slowly
	Record check valve 2 differential > or = 1.0 psi
Final	Close test cocks—remove all equipment
	Open all needle valves on test kit
	Open outlet shut-off valve slowly

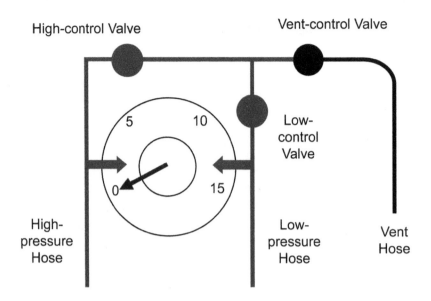

Figure 8-32   Three-valve test kit

Svb Field Test with Differential Pressure Gauge

Preparation	Notify customer; Inspect area; Remove canopy
	Flush test cock and vent valve
	Install fittings
	Inspect test kit—close all needle valves
Test Check Valve Test #1	Install test gauge and low pressure hose level with body. Attach compensating (bleed-off) tee to test cock
	Attach high hose to compensating tee; Open test cock slowly
	Open high bleed—bleed air from gauge; Close high bleed
	Close outlet shut-off valve—close inlet shut-off valve
	Open vent valve to lower pressure in body; When water stops running from vent valve, <u>record</u> value of check valve > or = 1.0 psi
Test air inlet valve Test #2	Place water on top of air inlet
	Open high bleed no more than ¼ turn
	<u>Record</u> when air inlet opens > or = 1.0 psi
Final	Close test cock and vent valve; Remove all equipment; Replace canopy
	Open inlet shut-off valve slowly
	Open outlet shut-off valve slowly. Open all needle valves on test kit

Copyright © 1/10/1990; rev. 01/12/2012; rev. 5/20/2013

TESTING PROCEDURES OR METHODS 179

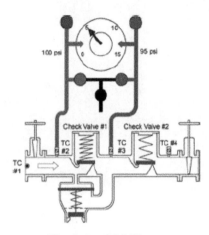
Test 1 - CV #1

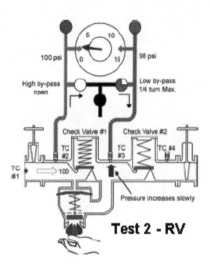
Test 2 - RV

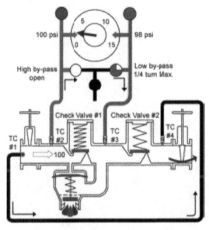
Test 2 - RV with by-pass

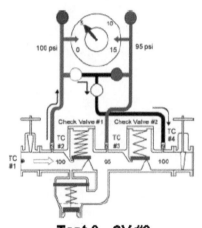
Test 3 - CV #2

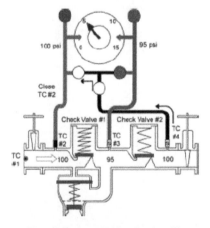
Test 5 - outlet shut-off

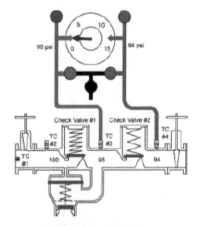
Test 6 - CV #2

Figure 8-33   Reduced-pressure field test with three-valve test kit

AWWA Manual M14

This page intentionally blank.

Appendix C

Industry Resources

NOTE: Abbreviations used refer to reference materials issued by the organization identified below:

ABPA	**American Backflow Prevention Association** 3016 Maloney Avenue, Bryan, TX 77801-4112 (979) 846-7606
AHAM	**Association of Home Appliance Manufacturers** 1111 19th St., N.W., Suite 402, Washington, DC 20036 (202) 872-5955
ANSI	**American National Standards Institute** 25 W. 43rd St., Fourth Floor, New York, NY 10036-7406 (212) 642-4900
ASME	**American Society of Mechanical Engineers** Two Park Ave., New York, NY 10016 (212) 591-7740
ASSE	**American Society of Sanitary Engineering** 18927 Hickory Creek Dr., Suite 220, Mokena, IL 60448 (708) 995-3019
ASTM	**American Society for Testing and Materials** 100 Barr Harbor Dr., West Conshohocken, PA 19428 (877) 909-2786
AWWA	**American Water Works Association** 6666 West Quincy Ave., Denver, CO 80235 (303) 794-7711
CISPI	**Cast Iron Soil Pipe Institute** 1064 Delaware Ave. SE, Atlanta, GA 30316 (404) 622-0073
CSA	**Canadian Standards Association International Etobicoke (Toronto)** 178 Rexdale Blvd., Toronto, ON, M9W 1R3, CANADA (800) 463-6727
CS&PS	**Commercial Standards and Product Standards** Superintendent of Documents, US Govt. Printing Office, 732 N. Capitol St., N.W., Washington, DC 20401

FM	**Factory Mutual Global** PO Box 7500, Johnston, RI 02919 (401) 275-3000
FS	**Federal Specifications** Superintendent of Documents, US Govt. Printing Office, 732 N. Capitol St., N.W., Washington, DC 20402
IAPMO	**International Association of Plumbing and Mechanical Officials** (Uniform Plumbing Codes) (UPC) 5001 E. Philadelphia St., Ontario, CA 91761-2816 (909) 472-4100
NFPA	**National Fire Protection Association** 1 Batterymarch Park, Quincy, MA 02169-7471 (800) 344-3555
NSF	**NSF International** 789 N. Dixboro Rd., PO Box 130140, Ann Arbor, MI 48113-0140 (800) NSF-MARK
PDI	**Plumbing and Drainage Institute** 800 Turnpike St., Suite 300, North Andover, MA 01845 (978) 557-0720; 1 (800) 589-8956
USC	**University of Southern California** **Foundation for Cross-Connection Control and Hydraulic Research** KAP-200 University Park MC-2531, Los Angeles, CA 90089-2531 (213) 740-2032
UL	**Underwriters' Laboratories Inc.** 333 Pfingsten Rd., Northbrook, IL 60062 (708) 272-8800
WQA	**Water Quality Association** 4151 Naperville Rd. Lisle, IL 60532-3696

Appendix D

Incidents Table

Description	Town Place	Date	Hazard Type	Link
A Commerce City family has been awarded nearly $1 million in damages after drinking water contaminated with raw sewage.	Denver, CO	Friday, October 26, 2012	Residential	http://denver.cbslocal.com/2012/10/26/commerce-city-family-awarded-nearly-1m-for-contaminated-water/
Bacteria known as coliform was discovered in the school's drinking water after a routine sample testing.	Malta, Idaho	Friday, September 28, 2012	School	http://minicassia-voice.com/featured/bacteria-found-in-raft-river-schools/
A cross-connection between an irrigation system and the culinary water system causes 13 families to get sick, some with *Giardia*.	Cedar Hills, Utah	Monday, September 17, 2012	Residential / Irrigation	http://www.abc4.com/content/news/top_stories/story/Tainted-water-made-Cedar-Hills-neighborhood-sick/19kK0Bikb0KnKCHGWSEcyQ.cspx
Private contractor mistakenly hooked up an incorrect pipe to another hook up.	Lake Perry, Miss.	Thursday, September 6, 2012	Residential	http://www.perryvillenews.com/latest_news/article_b8ca5e5a-fdb5-11e1-ba2f-0019bb30f31a.html
As many as 80 residents got sick over Memorial Day weekend, due to a nasty bacteria that got into the drinking water.	Boise, Idaho	Friday, June 8, 2012	Residential	http://boisestatepublicradio.org/post/boise-residents-sickened-bad-water-not-satisfied-companys-response
Andrea Mock reports a personal story of how her family home's water was contaminated.	Columbia, SC	Thursday, November 3, 2011	Residential / Irrigation	http://www.wltx.com/video/default.aspx?bctid=1257392162001

Description	Town Place	Date	Hazard Type	Link
Were alerted the water system might have been contaminated with water and air from a private residence.	Hickson, Ont.	Thursday, October 6, 2011	Residential / Well	http://www.woodstocksentinelreview.com/2011/10/06/boil-water-advisory-issued-for-hickson
Hydroseed material entered the town's water system through unauthorized connection to hydrant.	Somerset, Mass.	Thursday, July 21, 2011	Hydrant	http://www.heraldnews.com/news/x1009564725/3-arrested-in-connection-to-Somersets-contaminated-water-scare
An accidental backflow of the powerful herbicide Round Up into the city's water system during a spraying operation.	Kansas	Wednesday, June 29, 2011	Irrigation	http://www.kdcountry94.com/2011/06/29/round-up-backflow-in-sylvan-grove/
About 200,000 gallons contaminated water into the water distribution system.	Dorval, Que.	Friday, June 3, 2011	Airport	http://www.cbc.ca/news/canada/montreal/story/2011/06/03/dorval-water-advisory.html
Customer connecting a surface water irrigation line to the water system without an approved backflow preventer.	Kilauea, Hawaii	Friday, May 27, 2011	Irrigation	http://thegardenisland.com/news/local/unlawful-connection-caused-water-warning/article_1b1d4d3a-8902-11e0-9a48-001cc4c03286.html
Chemicals used in the cleaning process entered the potable water system of the school.	Oakville, Ont.	Wednesday, April 20, 2011	School	http://www.insidehalton.com/news/article/996874--region-working-with-school-after-water-contamination-leads-to-closure
Discolored water that came out of a custodial utility sink.	Chicago, Ill.	Monday, November 15, 2010	School	http://beaconnews.suntimes.com/news/2396109-423/batavia-system-testing-barshinger-bottled.html
Glycol, a chemical used in the school's heating and cooling system, leaked into the school's drinking supply.	Kentucky	Wednesday, October 13, 2010	School	http://www.bgdailynews.com/breaking/water-contamination-at-bristow-school/article_de8aaa1c-a9c3-5db3-8e07-bad405af27e8.html
After a company twice contaminated the water system at Normandale Community College last year.	Minneapolis, Minn.	Monday, October 11, 2010	School	http://www.startribune.com/printarticle/?id=104712149
74 students complained of nausea after drinking the water at the Flushing Elementary School.	Queens, N.Y.	Saturday, February 6, 2010	School	http://www.popularmechanics.com/home/improvement/electrical-plumbing/glycol-in-school-drinking-water
Residents told not to drink tap water after recycled water was mistakenly piped to them.	Coomera, Australia	Saturday, December 5, 2009	Residential	http://www.goldcoast.com.au/article/2009/12/05/166165_gold-coast-news.html

Description	Town Place	Date	Hazard Type	Link
Homeowner illegally connected two systems through their piping, and the secondary water was pushed through the line at a higher pressure, forcing backflow into the city waterline.	Layton, Utah	Thursday, October 15, 2009	Residential	http://www.watertechonline.com/articles/illegal-cross-connection-blamed-for-giardiasis
Fire trucks' water pressure overwhelmed the city's drinking supply lines and pushed fire suppression foam into them as firefighters tried to extinguish a burning Strip District warehouse.	Pittsburgh, Pa.	Tuesday, March 11, 2008	Hydrant	http://calfire.blogspot.ca/2008/03/us-news-firefighting-foam-contaminates.html
Pink-colored water coming from the taps in her home.	Stratford, Ont.	March 7, 2005	Car Wash	http://news.ontario.ca/archive/en/2006/06/23/Festival-U-Wash039-and-manager-fined-total-of-75000-Stratford.html

Other Websites for Backflow Incidents

http://www.abpa.org/incidents.htm

http://www.epa.gov/ogwdw/disinfection/tcr/pdfs/issuepaper_tcr_crossconnection-backflow.pdf

http://www.crd.bc.ca/water/crossconnection/incidentsandevents.htm

http://www.nobackflow.com/casehist.htm

http://www.bcwwa.org/cross-connection-control/251-backflow-incidents.html

NOTE: Links were valid as of date of publication of this manual.

This page intentionally blank.

Glossary

Absolute pressure The gauge pressure plus atmospheric pressure. It is measured in units of pounds per square inch absolute (psia).

Administrative authority An official office, board, department or agency authorized by law to administer and enforce regulation and or code requirements. This includes a duly authorized representative of the administrative authority.

Actual connection A cross-connection for which the connection exists at all times.

Air gap (AG) The unobstructed vertical distance through free atmosphere between the lowest effective opening from any pipe or faucet conveying water or waste to a tank, plumbing fixture, receptor, or other assembly and the flood level rim of the receptacle. These vertical, physical separations must be at least twice the effective opening of the water supply outlet, never less than 1 in. (25 mm) above the receiving vessel flood rim. Local codes, regulations, and special conditions may require more stringent requirements.

Air-gap fitting The physical device engineered to produce a proper air-gap separation as defined above.

Approved Accepted by the authority having jurisdiction as meeting an applicable standard, specification, requirement, or as suitable for the proposed use

Area protection The installation of a backflow prevention device on a section of piping system within a building or facility with both potable and non potable connections downstream of the device.

Assembly A backflow prevention device equipped with approved shut-off valves.

Atmospheric pressure The pressure exerted by the atmosphere at any point. Such pressure decreases as the elevation of the point above sea level increases. One atmosphere is equivalent to 14.7 psi (101.4 kPa), 29.92 in. (760 mm) of mercury, or 33.9 ft (10.1 m) of water column at average sea level.

Atmospheric vacuum breaker (AVB) A backflow prevention device consisting of a float check, a check seat, and an air-inlet port. A shutoff valve immediately upstream may or may not be an integral part of the device. The device is designed to allow atmospheric pressure to enter the air-inlet port thus breaking the vacuum and preventing backsiphonage. The device shall be installed at least six inches above the highest outlet and may never be subjected to a backpressure condition or have a downstream shutoff valve, or be installed where it will be in continuous operation for more than 12 hours.

Authority having jurisdiction The agency, organization, office, or individual responsible for approving materials, equipment, work, installation, or procedure.

Auxiliary water sources Any water source of water on or available to the premises other than the water supplier's approved source(s) of water. These auxiliary sources of water may include but not limited to other public water system sources or other unapproved onsite source(s) which are not under the control of a public water system, such as a well, lake, spring, river, stream, harbor, and so forth. Auxiliary water sources may also include graywater, rain or storm water, reclaimed waters, or recycled waters.

Backflow The reversal of flow of a liquid, gas, or other substance in a piping system.

Backflow preventer An assembly, device, or method that prevents backflow.

Backpressure A type of backflow where the pressure is higher than the incoming supply pressure.

Backsiphonage A type of backflow where the upstream pressure in a piping system is reduced to a subatmospheric pressure.

Ball valve A valve with a spherical gate providing a tight shutoff. Ball valves on backflow assemblies shall be fully ported and resilient seated.

Barometric loop A method of backflow prevention used to prevent backsiphonage only. It is a looped piping arrangement 35 ft (11 m) in height, It is based on the principle that water in a total vacuum will not rise higher than 33.9 feet at sea level.

Canadian backflow prevention assemblies (As defined in Candian Standards Association (CSA) Standard B64.

Air space type vacuum breaker (ASVB)
Atmospheric vacuum breaker (AVB)
Dual check with atmospheric port (DCAP)
Dual check with atmospheric port for carbonators (DCAPC)
Double check valve (DCVA)
Double check valve type for fire systems (DCVAF)
Dual check valve (DuC)
Dual check valve type for fire systems (DuCF)
Dual check valve type with intermediate vent (DuCV)
Hose connection dual check vacuum breakers (HCDVB)
Hose connection vacuum breaker (HCVB)
Laboratory faucet type vacuum breaker (LFVB)
Pressure vacuum breaker (PVB)
Reduced pressure principle type (RP)
Reduced pressure principle type for fire systems (RPF)
Single check valve type for fire systesm (SCVAF)
Spill-resistant pressure vacuum breaker (SRPVB)

Certified backflow prevention assembly tester A person who has demonstrated competence as evidenced by certification that is recognized by the approving authorityto field test, a backflow-prevention assemblies.

Consumer A person who utilizes water from a public water system.

Containment protection Backflow protection on the water pupply Line to a premises that is installed as close to the service connection to the public water system as possible. See also premises isolation.

Critical level A reference line representing the level of the check valve seat within a backsiphonage control unit. It is used to establish the height of the unit above the highest outlet or flood level rim. If it is not marked on the backflow preventer, the bottom of the assembly is the critical level.

Cross-connection An actual connection or a potential connection between any part of a potable water system and any other environment that would allow substances to enter the potable water system. Those substances could include gases, liquids, or solids, such as chemicals, water products, steam, water from other sources (potable or nonpotable), or any matter that may change the color or add odor to the water. Bypass arrangements, jumper connections, removable sections, swivel or change-over assemblies, or any other temporary or permanent connecting arrangement through which backflow may occur are considered to be cross-connections.

Cross-connection control A program to eliminate cross-connections or to prevent them from causing a public health threat.

Cross-connection control survey The review of the plumbing system to determine the existence of potential or actual cross-connections and to assess the degree of hazards of protected and unprotected cross-conneciotns.

Degree of hazard The assessment or evaluation of a facilities domestic water system's cross-connections as they relate to the health hazard of the consumers of water.

Direct cross-connection A cross-connection that is subject to both backsiphonage and backpressure.

Disk The part of a valve that actually closes off flow.

Double check valve assembly (DC or DCVA) A backflow prevention consisting of two internally loaded independently operating check valves, located between two tightly closing resilient-seated shutoff valves with four properly placed resilient-seated test cocks. This assembly shall only be used to protect against a non-health hazard (i.e., a pollutant).

Double check detector backflow prevention assembly (DCDA) A backflow prevention assembly specially designed with a line-size-approved double check valve assembly with a bypass containing a specific water meter and an approved double check valve assembly. The meter shall register accurately for only very low rates of flow up to 2 gpm and shall show a registration for all rates of flow. This assembly shall only be used to protect against a non-health hazard (i.e., a pollutant).

Effective opening The minimum cross-sectional area at the point of water supply discharge, measured or expressed in terms of the diameter of a circle, or if the opening is not circular, the diameter of a circle of equivalent cross-sectional area.

Electrolysis The corrosion resulting from the flow of electric current.

Expansion tank A tank used for safely controlling the expansion of water.

Field testing A procedure to determine the operational and functioning status of a testable backflow prevention assembly.

Fire department connection (FDC or Siamese connection) A connection through which a fire department can introduce supplemental water with or without the addition of other chemical fire-retarding agents by the is of a pump into a fire sprinkler system, standpipe, or other fire-suppression system.

Fire protection systems (water based)

Antifreeze system A wet-pipe sprinkler system containing antifreeze

Combined dry pipe-preaction system A sprinkler system containing air under pressure with a supplemental detection system installed in the area of the sprinklers. The detection system actuates tripping devices that open water inlet and air exhaust valves, which generally precedes the opening of the sprinklers. The detection system additionally serves as a fire alarm system.

Deluge system A sprinkler system having open sprinkler heads connected to a water supply. The sprinkler system piping is dry until the fire-detection system opens the water supply valve to the system.

Dry-pipe system A sprinkler system containing air or nitrogen under pressure and connected to a water supply. A sprinkler head opening allows the air or nitrogen to be released from the system and water to enter the system. Dry-pipe systems are to be maintained dry at all times. *Exception: During nonfreezing conditions, the system can be left wet if the only other option is to remove the system from service while waiting for parts or during repair activities.*

Foam water sprinkler and spray systems A special fire protection system pipe connected to a source of foam concentrate and to a water supply. The system may discharge the foam agent before, after, or with the water over the area to be protected.

Preaction system A sprinkler system containing air that may or may not be under pressure and connected to a water supply but having a supplemental detection system in the area of the sprinklers that would open a supply valve, allowing water to flow into the system and to be discharged by any open sprinkler.

Sprinkler system A system of underground and overhead piping hydraulically designed and constructed to which sprinkler heads are attached for extinguishing fire.

Standpipe system A piping system having valves, hose connections, and allied equipment installed within a premises, building, or structure where the hose connections are located in a manner to discharge water through an attached hose and nozzle to extinguish a fire. These systems may be wet or dry and may or may not be directly connected to a drinking water supply system. They may also be combined with a sprinkler system. There are three classes of standpipe systems. Class I service provides 2 1/2-in. hose stations from a standpipe or combined riser. Class II service provides 1 1/2-in. hose stations from a standpipe, combined riser, or sprinkler system. *Exception: A minimum 1-in. hose may be used for Class II Light Hazard Occupancies if investigated, listed, and authorized by the authority having jurisdiction.* Class III service provides 1 1/2-in. and 2 1/2-in. hose connections or 1 1/2-in. or 2 1/2-in. hose stations from a standpipe or combination riser.

Wet-pipe system A sprinkler system containing water and connected to a water supply.

NOTE: FOR FURTHER INFORMATION ON FIRE PROTECTION SYSTEMS, REFER TO NATIONAL FIRE PROTECTION ASSOCIATION PUBLICATIONS.

Flood level rim The level from which liquid in plumbing fixtures, appliances, or vats could overflow to the floor, when all drain and overflow openings built into the equipment are obstructed.

Gauge pressure The pressure at a point of a substance (gas or liquid) and is calculated by subtracting the atmosphere from the absolute pressure.

Graywater Untreated household used water, such as wash or rinse water from a sink, bathtub, or other household plumbing fixture, except a toilet, that does not contain animal wastes.

High health hazard (high hazard) A cross-connection or potential cross-connection involving any substance that could, if introduced into the potable water supply, cause death or illness, spread disease, or have a high probability of causing such effects. An example of substance would be any one of the National Primary Drinking Water Standards.

Indirect cross-connection A cross-connection that is subjected to backsiphonage only.

Individual protection The installation of a backflow protection device at the connection to a fixture or appliance where the cross-connection exists.

Inspection A visual examination of a backflow protection device or assembly, materials, workmanship or portion thereof to verify installation and operational performance of the device or assembly.

Inspector An individual working for or under the authority having jurisdiction empowered to ensure code compliance.

Internal protection Isolation, with respect to a potable water system, of a fixture, area, zone, or some combination thereof. Isolation at the fixture is installing an approved backflow preventer at the source of the potential contamination. Area or zone isolation entails confining the potential source within a specific area.

Irrigation water High hazardous water utilized for plant life.

Listed-classified-approved Materials, equipment, fixtures, and other products included in a list published by an agency or organization that has successfully evaluated the item and determined compliance with the agency's established material and/or performance standards.

Low health Hazard (low hazard) A cross-connection or potential cross-connection involving any contaminant that if introduced into the potable water system as a result of a backflow situation may cause a cosmetic effects (such as skin or tooth discoloration) or aesthetic effects (such as taste, odor, or color) in drinking water.

Maintenance Work performed or repairs made to keep equipment operable and in compliance.

Mushroom valve See *poppet valve*.

Needle valve A valve that has a small opening that is closed or opened by a needle-like spindle. Used for fine control.

Non-health hazard (low hazard) Any substance that generally would not be a health hazard but would constitute a nuisance or be aesthetically objectionable if introduced into the potable water supply such as the National Secondary Drinking Water Standards.

Plumbing system All potable water and distribution pipes, fixtures, traps, drainage pipe, gas pipe, water treating or using equipment, vent pipe, including joints, connections, devices, receptacles, and appurtenances within the property lines of a premises.

Pollution See *Nonhealth hazard*.

Poppet valve A valve consisting of a flat disk that raises and lowers without rotation about the valve opening. It is kept in position and on its path of travel by a rod or shaft attached to the disk at right angles to it and extending through the valve opening into a groove or hole that guides its movement. A poppet valve is also called a mushroom valve.

Potable water Water that is safe for human consumption as described by the public health authority having jurisdiction.

Potential cross-connection A connection for which something must be done to complete the cross-connection.

Premises isolation The prevention of backflow into a public water system form a user's premises by installation of a suitable backflow prevention device at the user's connection.

Pressure vacuum-breaker assembly (PVB) A backflow prevention assembly consisting of an independently operating, internally loaded check valve, an independently operating, loaded air-inlet valve located on the discharge side of the check valve, with properly located resilient-seated test cocks and tightly closing resilient-seated shutoff valves attached at each end of the assembly. This assembly is designed to be operated under pressure for prolonged periods of time to prevent backsiphonage. This assembly shall be installed at least twelve inches above the highest outlet. The pressure vacuum breaker may not be subjected to any backpressure and shall be tested at least annually and is suitable for high hazard cross-connections.

Public water system A system for the provision to the public of water for human consumption through pipes or other constructed conveyances, if such system has at least fifteen service connections or regularly serves at least twenty-five individuals for at least 60 days per year. EPA has defined three types of public water systems:

Community Water System (CWS) A public water system that supplies water to the same population year-round.

Non-Transient Non-Community Water System (NTNCWS) A public water system that regularly supplies water to at least 25 of the same people at least six months per year, but not year-round. Some examples are schools, factories, office buildings, and hospitals which have their own water systems.

Transient Non-Community Water System (TNCWS) A public water system that provides water in a place such as a gas station or campground where people do not remain for long periods of time.

USEPA also classifies water systems according to the number of people they serve:

Very Small water systems serve
25–500 people
Small water systems serve 500–3,300 people
Medium water systems serve
3,301–10,000 people
Large water systems serve
10,001–100,000 people
Very Large water systems serve
100,001+ people

Reclaimed water Water that, as a result of treatment of wastewater, is suitable for a direct beneficial use or a controlled use that would not otherwise occur and is not safe for human consumption.

Reduced-pressure principle backflow-prevention assembly (RP or RPBA or RPA or RPZ) A backflow prevention device assembly consisting of a mechanical, independently acting, hydraulically dependent relief valve, located between two independently operating, internally loaded check valves that are located between two tightly closing resilient-seated shutoff valves with four properly placed resilient-seated test cocks. This assembly shall be tested at least annually and is suitable for direct high hazard cross-connections.

Reduced-pressure principle detector backflow prevention assembly (RPDA) A specially designed backflow assembly composed of a line-size approved reduced-pressure principle backflow-prevention assembly with a bypass containing a specific water meter and an approved reduced-pressure principle backflow-prevention assembly. The meter shall register accurately for only very low rates of flow up to 2 gpm and shall show a registration for all rates of flow. This assembly shall be used to protect against a non-health hazard (i.e., a pollutant) or a health hazard (i.e., a contaminant). The RPDA is primarily used on fire sprinkler systems.

Service connection The connection between the public water system distribution system main and a user's domestic water system.

Service protection See *containment protection*.

Secondary protection The installation of a backflow prevention device or assembly on the public water system after the meter as close to the service connection to the public water system as possible. This installation is considered a redundancy backflow prevention device/assembly to augment the installation of the backflow prevention device and assemblies on the internal domestic plumbing system. This device shall not be installed in place of the internal prevention devices.

Sewage Liquid waste containing human, animal, chemical, or vegetable matter in suspension or solution.

Single check valve A single check valve is a directional flow control valve, but not an approved backflow preventer.

Spill resistant pressure vacuum breaker A testable vacuum breaker assembly containing a check valve and an air-inlet valve that must seal to the atmosphere, before water flows through the check valve.

Static water level The height measurement of a liquid at rest within a vessel.

Submerged inlet An inlet pipe opening that is below the flood level rim of the receptacle.

Test equipment An electronic or mechanical instrument recognized by the authority having jurisdiction to field-test the operational performance of a backflow preventer (*see* Field testing)

Union A three-part coupling device used to join pipe.

Valve seat The port(s) against or into which a disk or tapered stem is pressed or inserted into to shut down flow.

Velocity The speed of motion in a given direction.

Water supplier The owner or operator of a public water system.

Zone protection The installation of a backflow prevention device on a sections of a piping system within a building or facility with no potable connections downstream of a device.

Index

NOTE: *f.* indicates figure; *t.* indicates table.

ABPA (American Backflow Prevention Association), 181
Absolute pressure, 11–12, 187
Access covers, 125
Accuracy of test equipment, 122
Actual connections, 19, 187
Administrative authority, 187
Administrator, cross-connection control programs, 29
AG. *See* Air gap
Agencies with jurisdiction over water supply, 3, 4*f*
AHAM (Association of Home Appliance Manufacturers), 181
Air gap (AG)
   applications, 60*f*, 61*f*
   defined, 187
   in distribution system, 84
   for fire hydrants, 78
   for fire protection systems, 79
   general discussion, 59
   inventory of, 38
   on lavatory, 60*f*
   for marine installations, 81
   overview, 21
   for reclaimed water systems, 82
   symbol for, 43*f*
   on tank, 59*f*
   test reports, 38
   for water-hauling equipment, 84
Air-gap fitting
   defined, 187
   symbol for, 43*f*
Alarm check valves, 79–80
American Backflow Prevention Association (ABPA), 181
American National Standards Institute (ANSI), 181
American Society for Testing and Materials (ASTM), 181
American Society of Mechanical Engineers (ASME), 181
American Society of Sanitary Engineering (ASSE) testing procedures, 181
   DCDA/DCDF five-valve test, 160–163
   DCDA/DCDF three-valve test, 158–160
   DC/DCF five-valve field-test procedure, 149–150
   DC/DCF one hose test, 167–169
   DC/DCF three-valve field-test procedure, 147–148
   PVB field-test procedures, 151–154
   PVB one hose test, 169–170
   RPDA field-test procedure, 155–157
   RP/RPF five-valve field-test procedure, 145–146
   RP/RPF one hose test, 166–167
   RP/RPF three-valve field-test procedure, 143–144
   SVB testing, 163–166
American Water Works Association (AWWA), 181. *See also* AWWA Pacific Northwest Section testing procedures
Annual test notice, 97
ANSI (American National Standards Institute), 181
Antifreeze system, 78, 189
Approved, defined, 187
Approved backflow preventers, lists of, 37–38
Approved test procedures, 128–129
Area protection, 187
ASME (American Society of Mechanical Engineers), 181
ASSE. *See* American Society of Sanitary Engineering testing procedures
Assemblies. *See also specific assemblies by name*
   assessing the effectiveness of, 21–22
   defined, 187
   versus devices, 41–42
   maintenance of, 63–64
   parallel, installing for continuous water service, 62
   safety procedures for, 125–126
Association of Home Appliance Manufacturers (AHAM), 181
ASTM (American Society for Testing and Materials), 181
Atmospheric pressure, 11, 187
Atmospheric vacuum breaker (AVB)
   defined, 187
   general discussion, 45–46, 45*f*
   hazard protection and applications, 90*t*
   symbol for, 43*f*
Authority having jurisdiction, defined, 187
Auxiliary water sources, 76–77, 187
AVB. *See* Atmospheric vacuum breaker
AWWA. *See* American Water Works Association

AWWA Pacific Northwest Section testing procedures
    DCVA/DCDA testing, 133–134, 134t
    differential pressure gauge test kits, 138, 139f–142f
    overview, 120–121
    preliminary steps, 130
    PVBA testing, 134–135, 137t
    RPBA/RPDA testing, 130–132, 133t
    safety procedures, 125–128
    SVBA testing, 136–137, 137t
    tester ethics, 123–125
    tester responsibilities, 121–122
    test kits, 137–138
    test procedures, 128–129
    test report forms, 122, 123, 124f
    training and certification of testers, 121

Backflow, defined, 65, 187
Backflow incidents
    reporting, 8, 39–40
    table of, 183–185
Backflow preventers
    air gap, 59, 59f, 60f, 61f
    assemblies versus devices, 41–42
    AVB, 45–46, 45f
    backflow devices, 42
    in Canada, 22t
    common symbols used for backflow devices, 43f
    DCDA, 56–57, 57f
    DC valve backflow prevention assembly, 50–53, 51f, 52f
    defined, 187
    dual check, 42–43, 44, 44f
    dual check with atmospheric vent, 44, 45f
    effectiveness of, assessing, 21–22
    field testing, 37–38, 61–64
    for fire hydrants, 78
    for fire protection systems, 78
    HCVB, 46–47, 46f
    installation and ownership of, 36
    installation of, factors to consider, 68
    inventory of, 38
    maintenance of, 63–64
    overview, 41
    PVB assembly, 47–49, 48f
    repairing, 63–64
    RP, 53–56, 54f, 55f
    RPDA, 57–58, 58f
    service protection, 68
    SVB, 49–50, 50f
    test reports, 38
    in United States, 22t
Backflow prevention assembly testers
    certification, 34, 121
    ethics, 123–125
    field testing, 61–64
    quality assurance programs, 38
    responsibilities of, 64, 120–122
    safety procedures, 34–35, 125–128
    test report forms, 122, 123, 124f
    training, 33–34, 121
Backflow prevention equipment locations, 4f
Backflow prevention principles
    assemblies and devices, assessing effectiveness of, 21–22
    backpressure, 16–17
    backsiphonage, 13–16
    basic hydraulics, 11–13
    degrees of hazard, assessing, 18–20
    other risk factors, assessing, 20
Backpressure. *See also* Backflow preventers
    caused by carbon dioxide cylinder, 17f
    defined, 187
    general discussion, 16–17
    risk assessment, 19–20
Backsiphonage. *See also* Backflow preventers
    caused by reduced pressure on booster pump, 15f
    defined, 187
    due to high rate of water withdrawal, 14f
    general discussion, 13–16
    risk assessment, 20
Bacteria, 6–7
Ball valve, 187
Barometric loop, 12–13, 12f, 187–188
Booster pumps
    backpressure backflow caused by, 16
    backsiphonage due to reduced pressure in, 14, 15f
Building codes, 3
Building owners
    notifications of survey findings, 72
    permission for water shutoff, obtaining, 130
    responsibilities, 8, 66

Canada
    backflow preventers in, 22t
    backflow prevention assemblies, 188
    oversight of public water supply in, 2
Canadian Standards Association International Etobicoke (CSA), 181
Carbon dioxide cylinders, backpressure caused by, 16, 17f
Cast Iron Soil Pipe Institute (CISPI), 181
CCR (consumer confidence report), 30
Certified backflow-prevention assembly testers, 34, 63, 121, 188
Check valve
    safety procedures, 125–126
    symbol for, 43f
Chemical contaminants, 85
    general discussion, 7
    health hazards, 20
    irrigation systems, 80–81

overview, 5
Chlorination, 5
Chronic health effects, 5
CISPI (Cast Iron Soil Pipe Institute), 181
Citrobacter freundii, 6
Classified facilities, hazards in, 83
Closed facilities, hazards in, 83
Coliform bacteria, 6
Combined dry pipe-preaction system, 189
Commercial facilities
   educational materials for customers, 30–31
   hazards in, 77
   service protection, 89t
Commercial Standards and Product Standards (CS&PS), 181
Common-law doctrines, 8–9
Community Water System (CWS), 191
Compliance notifications, 72
Comprehensive programs, 26, 28
Computer programs for record keeping, 72–73
Conducting cross-connection control surveys, 69–70
Confined space entry, 126, 127
Consumer, defined, 188
Consumer confidence report (CCR), 30
Containment
   benefits of, 27
   general discussion, 25
   overview, 24
   for specific customers, 89t
   supplier-owned offices and work areas, 88
Containment protection
   cross-connection control surveys, 68
   DC/DCVA assemblies, 90t
   defined, 188
   degree of hazard, assessing, 67
   RP assemblies, 89t
Continuous water service, 62
Control of records, 33
Coordination with local authorities, 31–32
Coordinator, cross-connection control programs, 29
Cost of cross-connection control programs, 28
Critical level, defined, 188
Cross-connection control programs
   choosing, 27–28
   comprehensive programs, 26
   containment isolation, 25
   coordination with local authorities, 31–32
   cross-connection control plans, 29–31
   defined, 188
   developing, 24–25
   documentation, 33
   human resources, 33–35
   importance of, 5
   internal protection, 25–26
   joint programs, 26–27
   management programs, 28–32
   overview, 23
   performance goals, 31
   program administration, 35–40
   regulations, 3
   resources, commitment to, 28–29
   types of, 25–27
Cross-connection control surveys
   authority and responsibilities, 66
   conducting, 69–70
   considerations and concepts, 67–68
   containment compliance notification, 94
   dedicated water line approach, 68
   defined, 188
   degree of hazard, assessing, 67
   notification and enforcement, 72
   overview, 65–66
   purpose of, 66
   record keeping, 72–73
   request for internal cross-connection control information, 68–69, 95
   service protection approach, 68
   survey compliance notice, 93–94
   survey noncompliance notices, 95–96
   survey notice, 93
   survey reports, 73
   survey shut-off notice, 96
   tools for, 70–71
Cross connections
   contamination risk, assessing, 19
   defined, 1, 65, 188
Cross-connection specialists, safety procedures for, 125–128
CS&PS (Commercial Standards and Product Standards), 181
CSA (Canadian Standards Association International Etobicoke), 181
Customers
   backflow prevention assembly testers responsibilities to, 122
   communications, 29–30
   hazard protection for specific, 89t–91t
   keeping records of, 33
   ownership of backflow prevention assemblies, 36
CWS (Community Water System), 191

Damage, claims of, 9
DC. *See* Double check valve assembly
DCDA. *See* Double check detector backflow prevention assembly
DCF. *See* Double Check Fire Protection Backflow Prevention Assemblies
DCVA. *See* Double check valve assembly
Dedicated water line approach, 68
Degree of hazard. *See also* Hazards; Public health aspects of cross-connection control

cross-connection control surveys, 67
defined, 188
risk assessments, 18–20, 35
vulnerability assessments, 35–36
Deluge systems, 79, 189
Devices. *See also specific devices by name*
versus assemblies, 41–42
common symbols used for, 43f
Differential pressure gauge test kits. *See also* Five-valve differential test kits; *specific assemblies by name*; Three-valve differential test kits
general discussion, 138
hose connections, 139f–142f
Direct cross-connection, defined, 188
Discontinuing water service, 32
Disk, defined, 188
Distribution system
hazards in, 84–85
overview, 5
piping, bacteria growth due to poor condition of, 6–7
Dockside watering points, hazards in, 81–82
Documentation
incident reporting, 39–40
notifications of survey findings, 72
sample notices and letters, 93–98
test reports for backflow prevention assemblies, 38
Domestic water systems, cross-connection control surveys of, 71
Double check detector backflow prevention assembly (DCDA)
defined, 188
five-valve field-test procedure, 160–163
general discussion, 56–57, 57f
testing using differential pressure gauge test kits, 134t, 137t
three-valve field-test procedure, 158–160
Double Check Fire Protection Backflow Prevention Assemblies (DCF)
five-valve field-test procedure, 149–150
one hose tests, 167–169
three-valve field-test procedure, 147–148
Double check valve assembly (DC or DCVA)
containment protection, 90t
defined, 188
for fire protection systems, 79, 80
five-valve field-test procedure, 149–150
general discussion, 50–53, 51f
hazard protection and applications, 90t
one hose tests, 167–169
single hose field test with differential gauge, 171–172, 171f, 172f
symbol for, 43f
testing using differential pressure gauge test kits, 133–134, 134t, 137t
three-valve field-test procedure, 100–106, 147–148
typical applications, 52f
Double wall with leak protection (DWP) heat exchangers, 83
Double wall with no leak protection (DW) heat exchangers, 83
Dry-pipe system, 78, 79, 189
Dual check, 42–43, 44, 44f
Dual check with atmospheric vent, 44, 45f
DW (double wall with no leak protection) heat exchangers, 83
DWP (double wall with leak protection) heat exchangers, 83

Effective opening, defined, 189
Ejector or aspirator unit, symbol for, 43f
Electrical hazards, 126, 127f
Electrolysis, defined, 189
Electronic databases, 72–73
Elevated piping, backpressure from, 16
Emergency plans, 5
Enforcement policy, 32, 72
EPA. *See* United States Environmental Protection Agency
Ethics of backflow prevention assembly testers, 123–125
Expansion tank, defined, 189
Extensions, compliance, 72

Facilities
hazards in commercial, industrial, and institutional, 77
hazards in restricted, classified, or other closed, 83
requesting internal cross-connection control information for, 68–69
Factory Mutual Global (FM), 182
FDC (fire department connection), 189
Fecal matter, 81
Federal drinking water regulations, 2, 3
Fertilizers, 80–81, 84
Field testing. *See also* Differential pressure gauge test kits
of backflow prevention assemblies, 37–38
certification for, 63
continuous water service, 62
DCVA three valve procedure, 100–106
defined, 189
field-test equipment, 63–64
field-test report forms, 63
procedures, 62–63
PVB three-valve procedure, 106–110
on reclaimed-water line and equipment, 82
RPBA three-valve procedure, 110–116
SVB three-valve procedure, 116–119
testers, 61, 64

testing awareness, 62
test reports, 38
Field-test report forms, 63
Fire department connection (FDC or Siamese connection), 189
Fire hydrant hazards, 77–78
Fire marshals, 3
Fire protection systems (water based)
    cross-connection control surveys, 71
    defined, 189
    hazards, 78–80
    service protection, 89t
Five-valve differential test kits
    DCDA/DCDF testing, 160–163
    DC/DCF field-test procedure, 149–150
    PVB testing, 153–154
    RP/RPF field-test procedure, 145–146, 174
    SVB testing, 164–166
Fixture protection
    benefits of, 27–28
    degree of hazard, assessing, 67
    general discussion, 25–26
    overview, 24
Flood level rim, defined, 189
FM (Factory Mutual Global), 182
Foam water sprinkler and spray systems, 189
Freeze resistant sanitary yard hydrants, 91t
Funding, cross-connection control programs, 28

Gate valve, symbol for, 43f
Gauge pressure (psig), 11, 189
Giardia lamblia, 6
Government statutes and regulations, 7–8
Graywater, 190
Grounding, 126, 127f

Hazardous materials, safety procedures for, 126
Hazards. *See also* Degree of hazard
    assessments by water suppliers, 75–76
    auxiliary water, 76–77
    in commercial, industrial, and institutional facilities, 77
    in distribution system, 84–85
    dockside watering points, 81–82
    electrical, 126, 127f
    fire hydrants, 77–78
    fire protection systems, 78–80
    incidents tables, 183–185
    irrigation systems, 80–81
    marine facilities, 81–82
    posed by water suppliers, 84–88
    protection for specific customers, 89t–91t
    reclaimed water, 82
    residential water services, 83
    restricted, classified, or other closed facilities, 83
    solar domestic hot-water systems, 83–84
    supplier-owned offices and work areas, 87–88
    water-hauling equipment, 84
    water treatment plants, 85–87, 86t, 87f, 88f
HCVB. *See* Hose connection vacuum breaker assembly
Health aspects. *See* Public health aspects of cross-connection control
Health inspectors, 3
Herbicides, 81, 84
High health hazard (high hazard), 89t, 190
Hose connection vacuum breaker (HCVB)
    general discussion, 46–47, 46f
    for irrigation systems, 90t
    for marine installations, 81–82
Human resources
    certification and competencies, 34
    safety, 34–35
    training, 33–34, 121
Hydraulic grade line, 12
Hydraulics, 11–13
Hydropneumatic tanks, 17

IAPMO (International Association of Plumbing and Mechanical Officials), 182
Incidents
    reporting, 8, 39–40
    tables of, 183–185
Indirect cross-connection, 190
Individual protection, 190
Industrial facilities
    hazards in, 77
    service protection, 89t
Industry resources, 181–182
In-premise protection
    benefits of, 27–28
    general discussion, 25–26
    overview, 24
Inspection, defined, 190
Inspectors
    defined, 190
    safety procedures, 125–128
Installation and ownership of backflow prevention assemblies, 36
Installation of backflow preventers. *See specific backflow preventers by name*
Institutional facilities
    hazards in, 77
    service protection, 89t
Internal cross-connection control information, request for, 68–69, 95
Internal potable water systems, cross-connection control surveys of, 71
Internal protection
    benefits of, 27–28
    defined, 190
    degree of hazard, assessing, 67
    general discussion, 25–26

overview, 24
supplier-owned offices and work areas, 88
Internal zone isolation, 68
International Association of Plumbing and Mechanical Officials (IAPMO), 182
Inventory of backflow prevention assemblies, 38
Investigation of backflow incidents, 39–40
Irrigation systems
hazard protection, 90t–91t
hazards, 80–81
symbol for, 43f
Irrigation water, defined, 190

Joint programs, 26–27
Jumper cables, temporary, 125, 126

Legal aspects
common-law doctrines, 8–9
government statutes, regulations, and local controls, 7–8
Liability for impure water supply, 8–9
Liquid-to-liquid solar heat exchangers, 83
Listed-classified-approved, defined, 190
Lists of approved backflow preventers, 37–38
Local authorities, coordination with, 7–8, 31–32
Low health Hazard (low hazard), 89t, 190

Maintenance
of assemblies, 63–64
defined, 190
of test kits, 137–138
Management programs
coordination with local authorities, 31–32
cross-connection control plans, 29–31
overview, 28
performance goals, 31
resources, commitment to, 28–29
Mandatory service protection, 24–25
Manufacturers of test kits, 137–138
Marine facilities, hazards in, 81–82
Microbiological contaminants, 6–7, 20, 85

Military facilities, hazards in, 83
Mushroom valve, 190

National Fire Protection Association (NFPA), 182
Needle valve, 190
New England Water Works (NEWWA)
DCVA three valve field-test procedure, 100–106
overview, 99
PVB three-valve field-test procedure, 100–106
RPBA three-valve procedure, 110–116
SVBA three-valve field-test procedures, 116–119

NFPA (National Fire Protection Association), 182
Non-health hazard (low hazard), 190
Non-Transient Non-Community Water System (NTNCWS), 191
Notices
annual test notice, 97
containment compliance notification, 94
request for internal cross-connection control information, 95
survey compliance notice, 93–94
survey noncompliance notices, 95–96
survey notice, 72, 93
survey shut-off notice, 96
testing shut-off notice, 98
NSF International, 182
NTNCWS (Non-Transient Non-Community Water System), 191

Offices and work areas, hazards in supplier-owned, 87–88
One hose tests
for DC/DCF assemblies, 167–169
DCVA field test with differential gauge, 171–172, 171f, 172f
for PVB assemblies, 169–170
for RP/RPF assemblies, 166–167
Overhead assemblies, safety procedures for, 126
Owners
notifications of survey findings, 72
permission for water shutoff, obtaining, 130
responsibilities, 8, 66
Ownership of backflow prevention assemblies, 36

Parallel assemblies, 62
PDI (Plumbing and Drainage Institute), 182
Performance goals, cross-connection control programs, 31
Permission by owner for water shutoff, 130
Personal protective equipment, 126
Personnel, cross-connection control programs, 28–29
Pesticides, 81, 84
Physical hazards, 7
Piping
bacteria growth due to poor condition of, 6–7
elevated, backpressure from, 16
Plumbing and Drainage Institute (PDI), 182
Plumbing codes, 3, 4f, 5, 68
Plumbing system, defined, 190
Point of entry/service, 2
Pollution, defined, 190
Poppet valve, 190
Potable water, defined, 6, 190
Potential connections, 19
Potential cross-connection, 190

Preaction systems, 79, 189
Premises isolation
   benefits of, 27
   cross-connection control surveys, 68
   defined, 190
   degree of hazard, assessing, 67
   general discussion, 25
   overview, 24
Pressure, 11
Pressure-reducing valve, symbol for, 43f
Pressure relief valve, symbol for, 43f
Pressure vacuum breaker (PVB) assembly
   defined, 190–191
   five-valve field-test procedure, 153–154
   general discussion, 47–49, 48f
   for irrigation systems, 81, 90t
   one hose test, 169–170
   symbol for, 43f
   testing using differential pressure gauge test kits, 134–135, 137t, 173
   three-valve field-test procedure, 106–110, 151–152
Pressurized containers, backpressure caused by, 16–17, 17f
Professional groups, educational materials for, 30–31
Protection from hazards for specific customers, 89t–91t
psig (gauge pressure), 11, 189
Public education programs, 29–30
Public health aspects of cross-connection control
   barriers established for, 5–6
   chemical contaminants, 7
   microbiological contaminants, 6–7
   physical hazards, 7
Public service announcements, 30
Public water system (PWS)
   barriers established for, 5–6
   cross-connection control surveys, 71
   defined, 2, 191
Pumping system, backpressure backflow caused by, 17, 18f
PVB. *See* Pressure vacuum breaker assembly
PWS. *See* Public water system

Quality assurance programs, 38–39

Radio announcements, 30
Rate of water withdrawal, backsiphonage due to, 13, 14f
Recertification, 34
Reclaimed water
   defined, 191
   hazards, 82
Record keeping, 33, 72–73
Recycled water, hazards in, 82

Reduced-pressure principle backflow-prevention assembly (RP or RPBA or RPA or RPZ)
   containment protection, 89t
   defined, 191
   for fire hydrants, 78
   for fire protection systems, 79, 80
   five-valve field-test procedure, 145–146, 174
   general discussion, 53–56, 54f, 55f
   for irrigation systems, 90t
   for marine installations, 81
   one hose test, 166–167
   for reclaimed water systems, 82
   testing using differential pressure gauge test kits, 130–132, 133t, 137t
   three-valve field-test procedure, 110–116, 143–144, 177
   two-valve field-test procedure, 175, 176f
   for water-hauling equipment, 84
Reduced-pressure principle detector backflow prevention assembly (RPDA)
   defined, 191
   for fire protection systems, 79, 80
   general discussion, 57–58, 58f
   symbol for, 43f
   testing using differential pressure gauge test kits, 130–132, 133t, 137t
   three-valve field-test procedure, 155–157
Reduced-Pressure Principle Fire Protection Backflow Preventers (RPF)
   five-valve field-test procedure, 145–146
   one hose test, 166–167
   three-valve field-test procedure, 143–144
Repairing backflow prevention assemblies, 63–64
Reporting requirements, SDWA, 8
Request for internal cross-connection control information, 68–69, 95
Residential facilities
   fire-sprinkler systems, 80
   service protection, 89t
   water services, hazards in, 83
Resources, commitment to, 28–29
Responsibility of treating cross-connection, 66
Restricted facilities, hazards in, 83
Retention requirements for records, 72
Revised Total Coliform Rule (USEPA 1999), 65
Risk assessment records, 33, 35
RP. *See* Reduced-pressure principle backflow-prevention assembly
RPA. *See* Reduced-pressure principle backflow-prevention assembly
RPBA. *See* Reduced-pressure principle backflow-prevention assembly
RPDA. *See* Reduced-pressure principle detector backflow prevention assembly
RPF. *See* Reduced-Pressure Principle Fire Protection Backflow Preventers

RPZ. *See* Reduced-pressure principle backflow-prevention assembly

Safety
   concerns in cross-connection control program, 34–35
   procedures for backflow assembly testing, 125–128
   publications, 129t
Safety inspectors, 3
Safety supervisors, 125
SDWA. *See* United States Safe Drinking Water Act
Secondary protection, defined, 191
Service connection, defined, 191
Service protection
   benefits of, 27
   cross-connection control surveys, 68
   defined, 191
   general discussion, 25
   for high-hazard categories of customers, 24–25
   overview, 24
   for specific customers, 89t
   supplier-owned offices and work areas, 88
Sewage, 191
Sewer flushing water, 84
Ship fire pumps, flushing, 17, 18f
Siamese connection, defined, 189
Single check valve, defined, 191
Single-family residential customers, educational materials for, 29–30
Single wall with no leak protection (SW) heat exchangers, 83
Solar domestic hot-water systems
   hazards in, 83–84
   recommended protection for, 84t
SOP (standard operating procedures), for fire hydrant use, 77
Sources of supply, 5, 6
Spill-resistant vacuum breaker (SVB)
   defined, 191
   five-valve field-test procedure, 164–166
   general discussion, 49–50, 50f
   for irrigation systems, 90t
   testing using differential pressure gauge test kits, 136–137, 137t, 178
   three-valve field-test procedure, 116–119, 163–164
Sprinkler systems, 78–79, 80, 189
Stakeholder meetings, 29
Standard operating procedures (SOP), for fire hydrant use, 77
Standpipe systems, 189
State drinking water regulations, 2, 3
Static water level, 191
Storage, 5
Submerged inlet, 191

Survey noncompliance notices, 95–96
Survey reports, 73
Surveys. *See* Cross-connection control surveys
SVB. *See* Spill-resistant vacuum breaker
SW (single wall with no leak protection) heat exchangers, 83
System operators, 5

Technical groups, educational materials for, 30–31
Television announcements, 30
Temporary jumper cables, 125, 126
Terminating water service, 32
testers. *See* Backflow prevention assembly testers
Testing procedures and methods. *See also* American Society Of Sanitary Engineering testing procedures; AWWA Pacific Northwest Section testing procedures; *specific devices and assemblies by name*
   backflow prevention assemblies, 62
   general discussion, 37–38
   overview, 99
Testing shut-off notice, 98
Test kits. *See also* Five-valve differential test kits; Three-valve differential test kits
   accuracy of, 122
   defined, 192
   differential pressure gauge, 138, 139f–142f
   general discussion, 137
   maintenance, 138
   manufacturers of, 137–138
   overview, 121
Test procedures, approved, 128–129
Test report forms, 38, 122, 123, 124f
Thermal expansion, backpressure from, 16, 82
Three-valve differential test kits
   DC/DCF field-test procedure, 147–148, 158–160
   DCVA field-test procedure, 100–106
   PVB testing, 106–110, 153–154
   RPBA field-test procedure, 110–116
   RPDA testing, 155–157
   RP/RPF field-test procedure, 143–144, 177
   SVB field-test procedures, 116–119, 163–164
Thrust restraints, 126
TNCWS (Transient Non-Community Water System), 191
Tools
   for cross-connection control surveys, 70–71
   safety procedures, 125
Toxic chemicals, 7
Trade groups, educational materials for, 30–31
Traffic hazards, 127–128
Training staff, 33–34, 121
Transient Non-Community Water System (TNCWS), 191
Treatment techniques, 5

TREEO. *See* University of Florida Training, Research, and Education for Environmental Occupations
Two-valve differential test kits, RP field tests with, 175, 176*f*

Underwriters' Laboratories Inc. (UL), 182
Union, defined, 192
United States Environmental Protection Agency (USEPA), 2, 65, 66
United States Safe Drinking Water Act (SDWA), 2, 3, 8
University of Florida Training, Research, and Education for Environmental Occupations (TREEO)
    DCVA single hose field test with differential gauge, 171–172, 171*f*, 172*f*
    overview, 99
    PVB field test with differential pressure gauge, 173
    RP field test with five-valve test kit, 174
    RP field test with three-valve test kit, 177
    RP field test with two-valve test kit, 175, 176*f*
    SVB field test with differential pressure gauge, 178
University of Southern California Foundation for Cross-Connection Control and Hydraulic Research (USC), 182
USEPA. *See* United States Environmental Protection Agency

Vacuum, 12–13
Valve seat, 192

Velocity, defined, 192
Venturi effect, 13
Vulnerability assessments, 35–36

Waterborne disease pathogens, 6–7
Water-hauling equipment, hazards related to, 84
Water main breaks, backsiphonage due to, 14, 15*f*
Water Quality Association (WQA), 182
Water quality monitoring, 5
Water suppliers
    backflow prevention assembly testers responsibilities, 122
    defined, 192
    hazard assessments, 75–76
    hazards posed by, 84–88
    responsibility of treating cross-connection, 66
Water treatment plants
    cross-connection control, 87*f*
    hazards in, 85–87
    recommended protection at fixtures and equipment found in, 86*t*
    service-containment and area-isolation, 88*f*
Water withdrawal, backsiphonage due to high rate of, 13, 14*f*
Wet-pipe systems, 79, 189
Workplace safety, 128
WQA (Water Quality Association), 182
Written service agreements, 32

Zone isolation, 68
Zone protection, 192

AWWA Manuals

M1, *Principles of Water Rates, Fees, and Charges,* #30001

M2, *Instrumentation and Control,* #30002

M3, *Safety Management for Water Utilities,* #30003

M4, *Water Fluoridation Principles and Practices,* #30004

M5, *Water Utility Management,* #30005

M6, *Water Meters—Selection, Installation, Testing, and Maintenance,* #30006

M7, *Problem Organisms in Water: Identification and Treatment,* #30007

M9, *Concrete Pressure Pipe,* #30009

M11, *Steel Pipe—A Guide for Design and Installation,* #30011

M12, *Simplified Procedures for Water Examination,* #30012

M14, *Backflow Prevention and Cross-Connection Control: Recommended Practices,* #30014

M17, *Installation, Field Testing, and Maintenance of Fire Hydrants,* #30017

M19, *Emergency Planning for Water Utilities,* #30019

M20, *Water Chlorination/Chloramination Practices and Principles,* #30020

M21, *Groundwater,* #30021

M22, *Sizing Water Service Lines and Meters,* #30022

M23, *PVC Pipe—Design and Installation,* #30023

M24, *Planning for the Distribution of Reclaimed Water,* #30024

M25, *Flexible-Membrane Covers and Linings for Potable-Water Reservoirs,* #30025

M27, *External Corrosion Control for Infrastructure Sustainability,* #30027

M28, *Rehabilitation of Water Mains,* #30028

M29, *Water Utility Capital Financing,* #30029

M30, *Precoat Filtration,* #30030

M31, *Distribution System Requirements for Fire Protection,* #30031

M32, *Computer Modeling of Water Distribution Systems,* #30032

M33, *Flowmeters in Water Supply,* #30033

M36, *Water Audits and Loss Control Programs,* #30036

M37, *Operational Control of Coagulation and Filtration Processes,* #30037

M38, *Electrodialysis and Electrodialysis Reversal,* #30038

M41, *Ductile-Iron Pipe and Fittings,* #30041

M42, *Steel Water-Storage Tanks,* #30042

M44, *Distribution Valves: Selection, Installation, Field Testing, and Maintenance,* #30044

M45, *Fiberglass Pipe Design,* #30045

M46, *Reverse Osmosis and Nanofiltration,* #30046

M47, *Capital Project Delivery,* #30047

M48, *Waterborne Pathogens,* #30048

M49, *Butterfly Valves: Torque, Head Loss, and Cavitation Analysis,* #30049

M50, *Water Resources Planning,* #30050

M51, *Air-Release, Air/Vacuum, and Combination Air Valves,* #30051

M52, *Water Conservation Programs—A Planning Manual,* #30052

M53, *Microfiltration and Ultrafiltration Membranes for Drinking Water,* #30053

M54, *Developing Rates for Small Systems,* #30054

M55, *PE Pipe—Design and Installation,* #30055

M56, *Nitrification Prevention and Control in Drinking Water,* #30056

M57, *Algae: Source to Treatment,* #30057

M58, *Internal Corrosion Control in Water Distribution Systems,* #30058

M60, *Drought Preparedness and Response,* #30060

M61, *Desalination of Seawater,* #30061

M65, *On-Site Generation of Hypochlorite,* #30065

M66, *Cylinder and Valve Actuators and Controls—Design and Installation,* #30066